Society of Earth Scientists Series

Editor:

Satish C. Tripathi
Lucknow
India

AF786369

For further volumes:
http://www.springer.com/series/8785

Rajiv Sinha • Rasik Ravindra
Editors

Earth System Processes and Disaster Management

Springer

Editors
Rajiv Sinha
Department of Civil Engineering
IIT Kanpur
Kanpur (UP)
India

Rasik Ravindra
The Society of Earth
Sciences, Indira Nagar
National Center for Antarctic and Ocean
Lucknow
India

ISBN 978-3-662-52191-5 ISBN 978-3-642-28845-6 (eBook)
DOI 10.1007/978-3-642-28845-6
Springer Heidelberg New York Dordrecht London

Printed on acid-free paper

Springer is part of Springer Science+Business Media (www.springer.com)

Preface

The Earth is a unique planet not only in our Solar System, but as far as we know, in the accessible universe. It is not just the only planet we have – it is the only living planet we know, or may ever know. Thus, it is the most remarkable characteristic of our planet. Humans are now perhaps the most geologically significant species ever. With growing population, our species is depleting water and fuel resources, causing floods and droughts, producing wastes and "greenhouse" gases – disturbing the planet's dynamic equilibrium. Protecting the present biosphere for the advancement of human society is a responsibility of us all. Therefore, understanding the Earth's system and its functioning is a prime area of research. We need to integrate our knowledge of different spheres to fully understand the complexity of the Earth System. Modern Earth System Science research not only tends to encompass the interactions among land, life, and water but also includes insights into the functioning and stability of paleo-ecosystems, understanding of biodiversity dynamics over long timescales, and predicting future biosphere vitality. The dynamic Earth System is complex and undergoes constant adjustment. Earth scientists are in the forefront of detecting these changes, and finding ways of mitigating them before it is too late. Natural disasters have posed considerable challenge to scientists for making accurate geological predictions. There is an urgent need to make the Earth scientists working on various domains to sit together to address these societal issues.

The Society of Earth Scientists (SES) was constituted during the International Year of Planet Earth-2008 with the aim to bring earth scientists of all the domains at a single platform, promote multidisciplinary/multiinstitutional research and communicate with the society to educate them and to understand their needs. The SES organized a National Conference on "Earth system processes and disaster management." Selected papers addressed to the topic are published as Special Publication-1 under "The Society of Earth Scientists Series" which is instituted jointly with Springer. The SES requested Dr. Rajiv Sinha, Professor of Geosciences, Indian Institute of Technology, Kanpur, U.P., India, and Dr. Rasik Ravindra, Director, National Centre for Antarctic & Ocean Research, Goa, India, to edit this publication. We are thankful to both of them for taking all efforts in bringing out this quality publication. The SES expresses sincere gratitude to Dr. Shailesh Nayak, Secretary, Ministry of Earth Sciences, Government of India and Patron of SES for extending all help and guidance for the cause of earth scientists. The support received from Dr. Chris Bendall, Publishing Editor, Springer- Earth Sciences, Heidelberg, Germany, who is instrumental in instituting this series, is thankfully acknowledged. The SES also expresses its gratitude to earth scientists who contributed their research work for this volume.

I hope this book will prove to be a valuble treasure of knowledge.

Satish C. Tripathi
Series Editor & Honorary Secretary
The Society of Earth Scientists

Contents

About the Editors

Rajiv Sinha is Professor of Geosciences, Indian Institute of Technology, Kanpur. After completing his Ph.D. from the University of Cambridge, UK, in 1992, Rajiv Sinha joined IIT, Kanpur, in 1994 and since then has been pursuing research to understand the dynamics of large river systems and long- and short-term response of river systems to climate change due to natural as well as anthropogenic forcing. His work on dynamics and flood risk evaluation of the Himalayan Rivers has provided important insights on process-based understanding of these rivers, which have implications for sustainable river management. He is currently working on the eco-geomorphological approach for the Ganga River management as a part of a national project.

Sinha has applied an integrated approach of remote sensing–based geomorphology with stratigraphic and sedimentologic analysis of the Ganga basin supported by shallow subsurface drilling down to depths of ~50 m which has given a new dimension to Quaternary studies in this country. This has enabled Sinha and his coworkers to demonstrate the significant role of climatic forcing on stratigraphic development in the Gangatic plains of the Himalayan foreland basin. These studies have highlighted the role of variability in monsoonal dynamics in understanding river processes, basin-scale geomorphology, and sediment flux over 10^3–10^5 years timescales. He has also worked on the lacustrine sedimentology notably on the Sambhar lake in Rajasthan which resulted in climatic reconstructions of the Thar region for the past 30,000 years using mineralogical and geochemical proxies. His work on the Priyadarshini lake in the Antarctica as a part of the XVI Expedition to Antarctica also involved paleoclimatic reconstruction at Holocene timescale. He has published over 85 papers in the national and international journals which have been cited in most of the front-ranking journals in the subject area by several research groups across the globe.

Sinha is a member of editorial board of important international journals, namely, *Journal of Sedimentary Research* (SEPM), *Quaternary International* (Elsevier), *Earth Surface Processes and Landforms* (Wiley Interscience), *The Open Geology Journal* (Bentham) and *e-Journal Earth Science India*. Sinha has coedited several special issues of the leading national and international journals: *Current Science* (2002), *Geomorphology*, (2005), *Quaternary International* (2006), *Global and Planetary Change* (2009), *Quaternary International* (2010), *Palaeogeography, Palaeoclimatology, Palaeoecology* (ongoing). These special issues have significantly increased the visibility of the Indian fluvial research in amongst the international community.

Rasik Ravindra is Director, National Centre for Antarctica and Ocean Research, an autonomous research institute under the Ministry of Earth Sciences, Government of India. He has undertaken several field operations while with Geological survey of India, between 1971 and 2005, that include elucidation of structure, metamorphism and stratigraphy of Pre-Cambrian rocks of Rajasthan, mineral resource potential of Alwar district, Geological and mineral investigations in Bhutan Himalaya, and ground validation of aeromagnetic anomalies in Meghalaya state.

His association with Antarctica dates back to 1987, when he undertook geological and glaciological investigations in Icy Continent. He led the subsequent Indian Expedition to Antarctica in 1989–1991 and visited the icy continent again in 1996–1997 and 2003–2004 with assignments on Environmental management and selection of site for new Research Station. In 2007, he initiated scientific activities in Arctic and opened an Indian Research station – Himadri at Ny Alesund, Svalbard. New Indian Station Bharati in Antarctica is coming up under his supervision. In 2010, he led the first successful Indian Expedition to South Pole.

Ravindra is Vice President of Scientific Committee on Antarctic Research, Chairman of INSA Committee of SCAR, Current Chairman of Asian Forum on Polar Science, Chairman of national Chapter of IODP, and Indian Representative to international bodies such as ATCM, COMNAP, and IASC. He is also the Chairman of the Committee on Dynamics of Himalayan Glaciers and one of the Vice Presidents of IGU, and Chairman of The Society of Earth Scientists.

He has been supervising and coordinating Indian Polar Science and logistic activities at the national level since 2005 and has coedited special sections on Polar sciences in *Current Science* and *Journal of Geological Society of India*. He has been conferred with National Mineral Award-1990, Antarctic Award-2002, and Rajiv Gandhi Excellence Award-2011 for his contributions to advancement of Polar sciences in India.

Chapter 1
Introduction

Rajiv Sinha and Rasik Ravindra

1.1 Earth System Processes

In recent years, the geosciences research across the world has become much more broad-based encompassing the entire earth system namely land, atmosphere, hydrosphere and biosphere. The Earth System Science is a truly interdisciplinary science which requires input from various disciplines namely, physics (earth processes), chemistry (earth materials), biology (fossil records and paleobotany) and mathematics (numerical modeling of earth systems). Apart from the emphasis on understanding the exogenic and endogenic processes, we have learnt to adopt a more holistic view of the earth as a 'system' which is a more global approach to understand the interaction between different components of the system – clearly distinct from the classical approach of 'geology'. There is also a clear recognition of the fact that earth system science education and research is essential for the 'sustainable development' of the society. In India, our geosciences curriculum as well as research has responded least to these developments. For example, the issues which concern our society today are water, climate, hazards, and resource. All of these issues require a proper understanding of the earth systems, and therefore, a concerted effort to promote Earth System Science in this country is necessary both at research as well as education front.

Earth is also a 'dynamic' system which means that there are continuous changes in the system as reflected in various processes operating on the earth. Earth System processes are both familiar and mysterious to most of us; river floods, coastal

R. Sinha (✉)
Department of Civil Engineering, Indian Institute of Technology Kanpur, Kanpur 208016, UP, India
e-mail: rsinha@iitk.ac.in

R. Ravindra
National Centre for Antarctic and Ocean Research, Head Land Sada Vasco-da Gama, Goa, India
e-mail: rasik@ncaor.org

R. Sinha and R. Ravindra (eds.), *Earth System Processes and Disaster Management*, Society of Earth Scientists Series 1, DOI 10.1007/978-3-642-28845-6_1,

storms, earthquakes, and landslides are all 'natural' processes but they become a 'disaster' due to the impact they produce on the human society. One of the fundamental tenets of earth system science is to help decipher this greatest of all epics. Because the processes that create strata occur all around us, we can easily see and study them; because they leave a permanent record only in particular areas and over very long time scales, we must learn to think about the familiar in new ways. The present book attempts to touch upon some of these issues with particular relevance to India. The book is divided in four major sections namely, Earth Systems – cryosphere and hydrosphere, Earth's climate system and paleoclimate, natural hazards and catastrophies, and finally, natural resource management.

1.2 Earth Systems: Cryosphere and Hydrosphere

This section focuses on two important components of the earth systems – cryosphere and hydrosphere. The first paper of this section by *Raina* presents an overview of the Himalayan glaciers amidst the controversies of the IPCC (2007) report that these glaciers would disappear by 2035. The author argues that the retreat of the Himalayan glaciers have been rather episodic since 12–15 ka and that the glacier snout fluctuation is not influenced by one single parameter such as the climatic changes but several other parameters, for example, physiographic characteristics of the accumulation zone and valley slope, play a more dominant role. Data available at present are too limited to draw any major conclusions about the role of global warming on the glacier retreat.

The next two papers in the book deal with the temporal variability of the glaciers in the Laddakh Himalaya. *Ganjoo and Koul* report the asynchronous behavior of the glaciers and provide evidences that contradict the much-hyped idea of fast melting of glaciers in the Himalaya. In line with the first paper by Raina, the authors argue that changes in the geometry and morphology of the glaciers are not exclusively the result of change in climate but are influenced by various other factors such as terrain morphology and neotectonics. *Tangri et al.* have documented the surging glaciers of the Shyok valley in the Karakoram Himalaya characterized by cycles of advancing or retreating trends punctuated by quiescent periods. Using repetitive satellite images for the period 1975–2009 in a GIS framework, the authors show that the behavior of four groups of glaciers namely the Rimo group, Chong Kumdan, Kichik Kumdan and Aqtash glaciers is distinctly apart from each other although the causative factors are still not well understood.

The next paper in this section by *Thamban et al.* deals with the ice cores from Antarctica and focuses on the century-scale climatic variability using stable oxygen and hydrogen isotope parameters. Two ice cores, IND-25/26 and IND-22/B4, representing 100 and 470 years of record were used for the study and variations in $\delta^{18}O$ records on an inter-annual to decadal scale have been shown to be associated with changes in low and mid latitude climatic modes like the Southern Annular Mode (SAM) and El Niño Southern Oscillation (ENSO). The authors have also estimated surface air temperatures using the $\delta^{18}O$ profiles of two ice cores. Data presented by the authors reveal a significant warming by 0.6–1°C per century in the coastal areas of Dronning Maud Land, with greatly enhanced warming during the past several decades (~0.4°C per decade).

The last paper in this section by *Sinha et al.* highlights the new developments and challenges in river science across the globe and argues that river science should be promoted in India in a major way. Recognizing that almost all rivers on this globe have been 'disturbed' significantly due to prolonged human interference, there is an urgent need to initiate an integrated effort to develop a process-based understanding of river systems of India with a strong emphasis on hydrology-geomorphology-ecology linkages. The Himalayan rivers systems of India are among the world's large river systems (Tandon and Sinha 2007) and they constitute the most important source of freshwater in the region. However, many of these rivers face severe stress due to over-exploitation of water and pollution. In many parts of India, wrong river management practices have led to serious damages to the ecosystem and this situation need to be urgently corrected. A recent report released by the National Academy of Science, Washington has noted that some of the important areas in river science research where a large global effort should be directed and they include environmental flows and river restoration, sediment transport and geomorphology, and groundwater-surface water interactions (NAS 2007). A recent initiative on the Ganga River Basin Management Plan being developed by the consortium of IIT's is laudable but there is still a long way to go in terms of manpower development, river health and river futures and River Science offers the right platform.

1.3 Earth's Climate System and Paleoclimate

Earth's climate system is a highly integrated system, which requires the understanding of atmosphere, ocean and land systems and their interactions within and with societies and technologies. Changes in regional hydrology can result from either natural climate variability (e.g. decadal changes in atmospheric and oceanic circulation) or from anthropogenic factors, and accurate attribution is quite difficult. Further, high-resolution paleoclimatic data from different parts of the world are necessary to constraint the climate models and to improve their prediction capabilities. India-specific data on climate change and paleoclimatic fluctuations are quite fragmentary at this stage. This section has a set of three papers that present a record of paleoclimatic fluctuations from the Arctic and Arabian Sea using multiple proxies. Another paper documents the anthropogenic climate change and projected vulnerabilities in the Indian context.

Singh and Ravindra have documented the impact of climate change on lichen and moss communities in Ny-Ålesund, located on the west coast of Spitsbergen in the Arctic region. The authors have characterized the communities in barren areas, open plains, bird habitats and wetlands. It has been observed by the authors that climate change is beginning to cause large-scale changes in Ny-Ålesund. Increase in summer temperature and glacial retreat in particular have caused the exposure of fresh areas for colonisation and succession of new communities in these areas. Several species of lichens and mosses in the tundra are disappearing due to competition and dominance by Angiosperms. This trend is likely to continue which may result in the replacement of ~40% of the population of lichens and mosses by flowering plants triggering a major change in the ecosystem and carbon cycle in this region.

Tiwari presents a high-resolution record of southwest monsoon for the last 2,800 years using wind-induced, upwelling related productivity records from the Arabian Sea. A sediment core (SK145-9) from the eastern Arabian Sea was analyzed for $\delta^{18}O$ and $\delta^{13}C$ variations of three different species of planktic foraminifera (*Globigerinoides ruber*, *Globigerinoides sacculifer* & *Globorotalia menardii*) to reconstruct paleo-precipitation records from this region. During the last 2,800 years, the authors have recorded five periods of aridity at 2,000, 1,500, 1,100, 850 and 500 year BP (calibrated ages) in the eastern Arabian Sea. This interpretation correlates well with wind intensity data from western Arabian Sea – reduced precipitation is accompanied by weakened winds. The author concludes that the SW monsoon precipitation governs the wind strength *via* latent heat release. Further, this study shows that carbon isotope variability ($\delta^{13}C$) of foraminifera in eastern Arabian Sea is an excellent indicator of productivity, which in turn is governed not only by wind induced mixing but also by the nutrient influx *via* surface runoff.

Patil and Singh have used mineral magnetism as a proxy to decipher paleoclimatic fluctuations in the eastern Arabian Sea. Various mineral magnetism parameters were generated for sediment samples obtained from a gravity core off Goa. Based on the variations in magnetic properties, the authors have identified three distinct periods (30–14 ka, 13.5–10.5 ka, and 9–5.5 ka) which have been correlated to major climatic events. The highest concentration of ferromagnetic minerals is observed during 13.5–10.5 ka which has been linked to arid conditions during the Younger Drayas event. The authors report no significant increase in magnetic mineral concentration since the last 10 ka except for a moderate increase during 9.5–8 ka. Higher values of magnetic parameters during 9–5.5 ka have been linked to high fluvial discharges during early Holocene monsoonal intensification.

Agnihotri and Dutta provide an overview of the available data on the anthropogenic climate change and discuss the vulnerabilities of natural systems towards the expected changes in temperature and precipitation. The decrease on the Eurasian snow cover, south Asian atmospheric pollution (aerosols), increasing frequency of extreme events, melting glaciers and rising sea levels and human health are some of the observed impacts and most of these have serious implications for human societies and eco-system. The authors argue that the anthropogenic drivers of climate change mutually interact among themselves and most importantly with natural drivers in terms of their net climatic impact. However, the separation of anthropogenic forcings from the natural factors remains a major challenge in climate research.

1.4 Natural Disasters

Several earth system processes such as earthquakes, volcanoes and floods turn into a disaster primarily due to the intensity of the process and the suddenness at which they occur. Most of these are natural processes but a high population density in such areas and lack of understanding of these processes generally result into severe loss of life and property. To top it all, incorrect management strategies make the situation worse and slow down the process of recovery. Three papers are presented in this section to cover a wide range of disasters such as landslides, earthquakes, and floods including the engineering principles in preventing the hazards related to tunneling.

Chakraborty and Anbalagan document a case study on landslide hazard from Nainital in the northwestern Himalaya. The authors have carried out a detailed slope stability analysis to understand the type of instability and slope characteristics. The area is located in an active tectonic belt and landslides are quite frequent in this region. Two most common modes of slope failure as identified by the authors include toppling failure and circular failure and these are primarily governed by the composition of slope material and type of instability. Suitable remedial measures for landslide hazard mitigation and management have also been suggested for these different types of failures for restoration of geo-environmental equilibrium of the area. *Kumar* provides an overview of the natural hazards of the Arabian Peninsula picking examples of geological (earthquakes, tsunamis and tidal waves, and volcanism), atmospheric (tropical cyclones and storm surges, sand/dust storms) as well as anthropogenic (droughts and desertification, and flash floods) hazards. The author provides an extensive documentation of these hazards in the Arabian Peninsula over historical time scale and argues for a comprehensive regional disaster management and prevention policy in the region. *Verma and Singh* have discussed the influence of Young's modulus and Poisson's ratio on the displacement around a circular tunnel. The authors examine the failure behavior near the tunnel face based on volumetric strain around tunnel and its behavior near the tunnel face and provide important insights for preventing any hazard in areas of tunneling.

1.5 Natural Resource Management

The first paper in this section by *Sain* deals with a viable future major energy source – gas hydrates- and examines the available data for evaluating the resource potential. Gas hydrates have attracted the global attention due to their huge energy potential but there are still large gaps in technology development to exploit this resource. The author has not only proposed several approaches for the identification and quantification of gas hydrate, but has also shown that there are possibilities of widespread occurrences of gas hydrates in the Krishna, Godavari, Mahanadi and Andaman regions in the Bay of Bengal. These findings are based on the anomalous reflector identified on seismic sections known as Bottom Simulating Reflector (BSR). The second paper in this section by *Pati et al.* review the research on impact craters and argues for intensive research to discover more impact structures from the Indian landmass. It is well established that these impact structures help in the generation and remobilization of the ore-bearing fluids and consequent mineralization. The fact that nearly 60% of the terrestrial impact structures are associated with some form of economic mineral deposits (Au, Ni-PGE, and U) including hydrocarbon deposits emphasizes the need to promote this research. The authors describe some of the major impact craters such as Dhala (Madhya Pradesh), Lonar (Maharashtra), and Ramgarh (Rajasthan), and provide an up-to-date status of research on impact structures in India.

The final paper of this book by *Motwani et al.* report in-situ measurements of spectral properties of the coastal waters of the Arabian Sea to understand the backscattering by total suspended matter. The authors have also attempted to correlate the backscattering coefficients with chlorophyll which has important implications for understanding the dynamics of marine particles and dissolved organic carbon. Since chlorophyll is a function of phytoplankton, these correlations can be useful for developing novel primary productivity models and also in estimating phytoplankton growth rate, growth conditions and bio-mass.

We sincerely hope that this book would be useful to improve our knowledge of the earth system processes particularly in the Indian context. We sincerely thank all authors of this special publication for their contribution.

References

IPCC [Intergovernmental Panel on Climate Change] (2007). Climate change 2007: synthesis report. Contribution of working groups I, II and III to the fourth assessment report of the intergovernmental panel on climate change [Core Writing Team, Pachauri, R.K]

NAS (2007) River science at the United States Geological Survey, 2010. National Acadamic Press, Washington, DC, p 206

Tandon SK, Sinha R (2007) Geology of large river systems. In: Gupta A (ed) Large rivers. Wiley, Chichester/England, UK

Section I
Earth Systems: Cryosphere and Hydrosphere

Chapter 2
Global Warming and the Glacier Retreat: An Overview

V.K. Raina

2.1 Introduction

Glacier ice cover in the Himalayas during the last Ice Age is believed to have been at least three times than what it is today. It started dwindling, with more and more area getting exposed from under the ice cover with each successive Interglacial warm Period. From whatever little we know about the Quaternary Ice cover in the Himalayas, it would appear that during the last glacial maximum Ice/glacier cover in the Himalayas had extended to around 20 odd kilometres downstream of the present glacier snout positions. Glaciers, for whatever the reason, stayed put around these altitudes till up to the end of LGM. Field evidences, pending proper dating data, enable us to correlate this position with the Kulti stage (Owen et al. 2001) of Lahul valley glaciation around 12–15 ka BP. Current phase of the glacier retreat in the Himalayas thus began around this time and has continued since then although episodically. There have been periods of relative halt preceded and succeeded by rapid withdrawal and production of significant glacier melt, obviously in response to climatic oscillations. As a matter of fact, the glaciers/glacier ice cover in the Himalayas has recorded considerable shrinkage both in area and volume (Raina and Srivastava 2008) and such observations, with rare exceptions here and there, have also been recorded since the monitoring of the Himalayan glaciers began around 150 years back (Raina 2006).

The science of glaciers is so complex and complicated that even today, it is not understood why one glacier, sometimes even within the same climatic zone, responds entirely differently to the environmental changes. For example, the Sonapani glacier in the Lahul valley of the Central Himalayas has retreated by slightly more than 1,000 m along the front (Fig. 2.1) during the last 100 years;

V.K. Raina (✉)
Geological Survey of India (Retired), 258, Sector 17, Panchkula 134109, Haryana, India
e-mail: vijay.raina5@gmail.com

R. Sinha and R. Ravindra (eds.), *Earth System Processes and Disaster Management*, Society of Earth Scientists Series 1, DOI 10.1007/978-3-642-28845-6_2,

Fig. 2.1 Relative positions of the snout of the Sonapani glacier, based on available data, in relation with terminal moraine ridge 'A'

whereas the Kangriz glacier in the Suru valley of the North West Himalayas (Fig. 2.2) has literally remained stationary during the same period.

It is true that the regime-health of a glacier depends primarily on two meteorological parameters: (a) snow precipitation – addition of mass and (b) atmospheric temperature – depletion of mass. As such one would expect all glaciers to response in a similar way to any climatic oscillation. Historical documentation of the glacier snout fluctuation for the last 150 odd years has revealed that this retreat/glacier snout fluctuation can be broadly regulated as (a) periodic, (b) accidental and (c) surging. The surging glacier movement is a separate entity that, atleast, at the present level of understanding, is not related to any climatic oscillation. The problem of resolving glacier movement as periodic or accidental is often extremely complicated. Unless and until the basic difference between the types of the glacier snout movement is understood, it may be erroneous to correlate glacier snout fluctuation with one atmospheric parameter or the other. More glaciers we can examine in a given area, the better are the chances of arriving at a correct conclusion.

Impact of global warming on the glacier/ice cover in the Himalayas has been an agenda for discussion at several national and international forums. Perception among the general public is that the global warming is likely to lead to rapid melting of the ice cover in the Himalayas, which could have devastating effects. While one may not doubt the fact that the climate, by and large, does appear to be getting warmer, what, however, does tax the mind is the attempted linkage of glacier retreat in the Himalayas to the global warming which is debatable and needs intensive research. Two important questions that need to be answered are: (a) are the glaciers in the Himalayas retreating at an abnormal rate, and (b) do the glaciers respond to immediate climatic change?

Fig. 2.2 Relative positions of the snout of the Kangriz glacier, Suru valley, over the last 100 years (Photo 1907: Neve; Photo 2000 (Anon) and Photo 2007: R.K. Ganjoo)

2.2 Glacier Retreat in the Himalaya

Historical retreat of the glaciers in the Himalayas based on the data generated over the last 150 years or so, is more or less constant with and in continuance of the secular retreat that began at the end of the last glacial cycle and like wise has been episodic. Periods of rapid and slow retreat alternate with each other with occasional stationary periods. At no time has the retreat or the area vacation from under the glacier ice cover been what can be termed as alarming/abnormal, especially in the last decade or so. To illustrate the point, examples of three glaciers of varying dimensions, from different physiographic zones of the Himalayas, and having an observation record of almost 100 years, has been sited below to enable the readers to arrive at a proper conclusion regarding the glacier retreat. Although a few glaciers can not be taken as the representative of around 10,000 glaciers of various sizes that exist in the Indian part of the Himalayas, however, these glaciers do represent the basic trend of glacier snout fluctuations. These glaciers have been in news in recent years with the reports that they are undergoing an abnormal retreat due to global warming, and are likely to vanish in next 50 years or so.

2.2.1 Siachen Glacier (IN 5Q 131084)*

The 74 km long Siachen glacier is located between the latitude 35° 11′ 20″ and 35° 43′ 30″N and longitude 76° 45′ and 77° 17′ 30″E covered under the Survey of India sheet number 52E.[1] This glacier, in the real sense of the term, is more than a valley glacier. It is a glacier system and should be referred as 'Siachen glacier system'. The Siachen glacier and its tributaries cover almost 1,500 km^2 of an area, and hence to use a term 'glacier' hardly suffices to convey to the reader its vast and immense physiographic impact.

The Siachen glacier, by and large, has not moved much during the last 50 years with relatively very little degeneration and vacation along the glacier front (Raina and Sangewar 2007). A comparison of the snout position with the help of photographs, taken at different times over the last 50 years from 1958 to 2011, does indicate vacation of the glacier snout along the centre. Glacier ice that used to extend as a thin tapering tongue along the west of centre has since melted away leaving behind a thick layer of terminal/surface moraines (Fig. 2.3). In 1958, there were two ice caves along the eastern valley wall separated by intervening area of the glacier ice that served as a tunnel. Ice tunnel has since been eroded as a result of the river erosion and there has been further retreat of the ice cave northwards.

[1] As per the convention of the International working group all glaciers under the inventory programme are numbered (IN stands for India, 5Q for Indus basin, other digits represent glacier classification and number (Sangewar and Shukla 2009)).

Fig. 2.3 (**a**) Siachen glacier showing tapering front along the west of centre in 1958 AD in relation to small gullies marked 'A' and 'B' (Photo: 1958, V.K. Raina). (**b**) Same view, 52 years later in 2010 AD, shows that the ice front has since melted away leaving behind remnant terminal moraines (Photo: 2010, Navin Juyal)

As a consequence, the glacier along the eastern limit shows a considerable degenerated and eroded condition as compared to its position in 1958 (Fig. 2.4). This erosion is more accidental than in response to any climatic change.

2.2.2 Gangotri Glacier (IN 50 131 06 029)

The Gangotri glacier located at longitude 30° 56′N and latitude 79° 04′E is about 31 km long covering an area of about 110 km^2. It is situated in the Bhagirathi valley of Uttarakhand, India, and is the largest glacier in the Central Himalaya. The snout of the Gangotri glacier, as is typical of glaciers in the Himalaya, is marked by a prominent ice cave (Fig. 2.5a) known as Gaumukh (means the mouth of a cow)

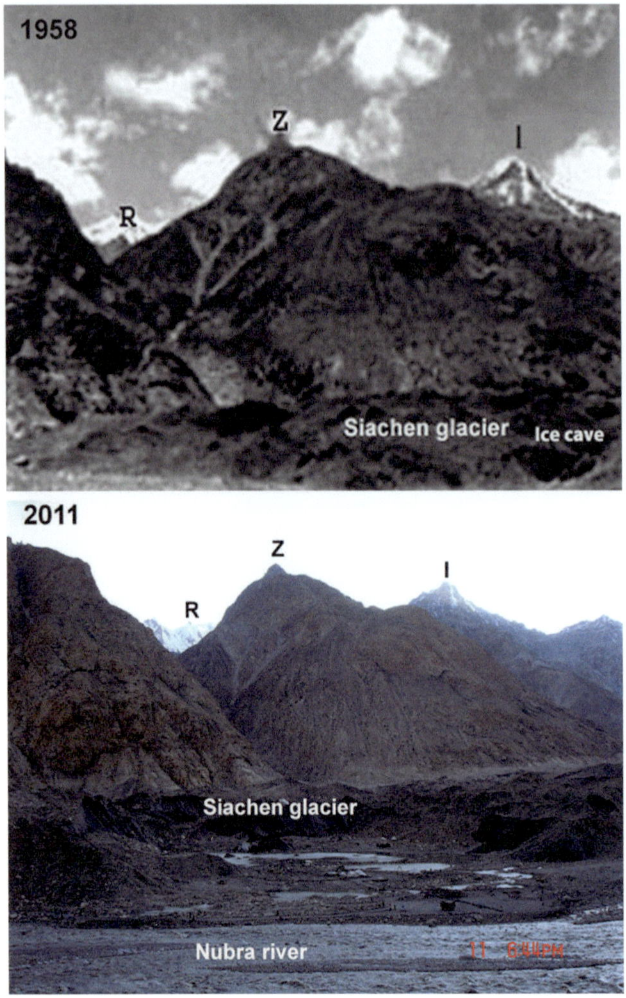

Fig. 2.4 Nubra river was flowing through an ice tunnel, within the glacier ice, along the eastern flank of Siachen glacier in 1958 (Photo: V.K. Raina). The tunnel has since collapsed and the glacier front eroded by the river water (Photo: SASE 2011)

from where the Bhagirathi river, one of the main tributaries of the river Ganges, originates. It has been the connoisseur of the explorers and the religious pilgrims over the centuries and has been photographed at regular intervals from 1860 onwards.

The Gangotri glacier, in its secular retreat, over last 75 years, for which the data is available, has shown a periodicity of around 35 years or so when it slows down considerably in its retreat and stays put in one place literally in a rest mode. The intervening periods are marked by rapid retreat (Table 2.1). The Gangotri glacier

Fig. 2.5 (**a**) Position of the Snout of Gangotri glacier in September 2004, 2009 and 2011 in reference to micro glacier landforms (indicated by *arrows*). (Photo 2004: Ishwar, Photo 2009: Kirit Kumar and Photo 2011: Tula Ram). (**b**) The Raktavaran stream (RS) has eroded a flow channel over the glacier. (**c**) Snout from 2009 onwards shows substantial degeneration along the eastern (*right*) side due to the development of a drain out for the Raktavran nala-accidental impact – that is likely to lead in breaking up of the glacier front as dead ice along the eastern side

Table 2.1 Area vacated by the Gangotri glacier along the snout front

Period	Area vacated annually by the glacier along its front in m^2/year	Based on
1935–1936	No change	GSI published data
1935–1956	2,500	
1956–1971	10,032	
1971–1974	594 (Practically no change-calving along the Gaumukh (ice cave) flanks)	
1974–1976	13,975	
1976–1977	8,900	
1977–1990	15,096	
1990–1996	22,370	
1968–2006	10,000 (0.01 $km^2\ a^{-1}$)	Bhambri et al. (2011)
2006–2007	Less than 5,000 (8.3 m)	Milap Sharma of JNU
2007–2008	No change	

had been showing a rather rapid retreat in the post-1974 period at an average of around 20 m per year until 2000 AD that led to the imaginative prediction of the vanishing of this glacier in next 35 years or so. In fact, since 2001 AD, the rate of the retreat has come down considerably and between September 2007 and June 2009, this glacier has been practically stationary. The Raktavaran melt water stream, which used to flow partially in a lateral valley and partially under the glacier ice, has since incised to form a deep channel (Fig. 2.5b) over the glacier surface. Figure 2.5c shows the snout on which substantial degeneration along the eastern (right) side due to the development of a drain out for the Raktavran nala can be observed. This is likely to result in breaking up of the glacier front (accidental impact) and 'dead ice' along the eastern side.

2.2.3 Shishram Nag/Shishram Glacier (5Q214<u>10</u> 029)

The East Lidar river has been generally described in geography books to be originating from a glacier tarn (a high altitude lake formed as a result of glacier action) known as Shishram Nag (34° 6′N: 75° 30′E), world famous for its scenic beauty and religious significance as it lies en-route to the sacred shrine of the Amarnath Cave. The glacier, at its head is the Shishram glacier, perhaps the largest glacier in the entire East Lidar valley. The Shishram glacier also marks the north facing slopes of the beautiful Kunyirhayan (Kohinoor) or Trinity peak (5,100 m). The glacier is slightly less than 3 km long with an overall area of about 6.25 km^2. In its upper reaches, the glacier is almost 2 km wide branching into two separated by a nunatak of a rocky promontory. The eastern branch glides down steeply in the form of a less than 500 m wide narrow stretch that used to culminate into a large bulbous front. It ranges in altitude from 4,900 m at its head to 3,900 m at the snout front. The western branch is less wide and much less steep.

Complete photographic record of the Shishram glacier is available since 1873 on the basis which one can say that the glacier has undergone major changes along the snout front. The glacier in the recent years, as can be assessed from the various photographs has not undergone any major change so far as the accumulation zone is concerned with the Equilibrium line being at and around 4,400 m. There, however, appears to have been a rapid degeneration of the glacier along the snout front followed by substantive withdrawal more so in last 2 years (2009–2011). Not only has the bulbous front of 1974 (Fig. 2.6) completely disappeared, the elongated snout front that used to override and extend below the rock knob has shrunk laterally eastward and its lower limit has thinned out being partially obscured by a snow bed as seen in the 2011 photograph (Fig. 2.6).

It is intriguing and interesting to note that while the glacier in September 2011 was exhibiting degeneration and crevassing along the snout front, it was still

Fig. 2.6 Relative position of the snout of the Shishramnag glacier, East Lidar valley, over the last 90 years (Photo 1921: Grinlinton, Photo 1974, 1982 and 2011: V.K. Raina; Photo 2008: Anilendra Bhhat, Internet)

covered by a snow bed at a considerable distance down stream of the snout. The tributaries were also snow covered on both side and a similar situation was noted at many places in the East Lidar from Shishram Nag Lake to Piush gorge over a distance of 8 km. If atmospheric temperature alone had the main role to control glacier melting then the snow bed downstream of the snout should have melted first.

Most of the glaciers in the Himalayas have shown periodic fluctuation along the glacier snout varying from degeneration in the early twentieth century with slow retreat, accumulation and mass growth in 1970s, slow upward movement of the equilibrium line towards the end of twentieth century followed by rapid degeneration along the glacier front in the last 2 years with slight increase/movement downwards of the equilibrium line.

2.3 Impact of Global Warming and Response of the Glaciers

Based on the observations of the glaciers for more than five decades, I have no hesitation in making a statement that a glacier does not necessarily respond to the immediate climatic changes. It was around the beginning of the twentieth century that the International community realised, for the first time, the possible impact of the rise in atmospheric temperature on mountain glaciers. A concerted effort world over including the Himalayas was initiated to identify and examine the fluctuations that the glaciers showed along the front-snout, which has continued since. It was believed that such a study over next 100 years or so would enable the scientists to establish the relationship between the climate change and the glacier fluctuations.

Glaciologists of the early twentieth century were handicapped by the fact that very little information was at hand about the accumulation zone of a glacier, as compared to present time, when the satellite imagery reveals all. Not only that, in some cases, rather in most of the cases, unless surveyed fully, one did not even have the knowledge as to how many tributaries were feeding the glacier. Despite this, a general conception, in those days, had been arrived at, that by recording the secular movements of glacier snouts, a measure of secular climatic change, either of snowfall, or of temperature, or of both could be derived at. Changes in the glacier snout positions would help in deciding, whether there is a definite recession of the snow-line; whether the climate is getting warmer or colder; a perception that, in one way or the other, has been persisting since.

The problem, however, is not as simple. A glacier is not only effected by complicity of the physical features that affect the glacier itself, but there are so many complexities of climate system that it is surprising that the snout movement of any glacier should reveal the periodic climate variation or the phases of the climatic variations till many centuries of observations become available. These complicated factors affect the periodic movements of the snouts of glaciers in a very marked manner. Ultimately the movements are due to climate and snowfall in particular, but the factors are so varied that the snout movements appear to be peculiar to each particular glacier. There may be little resemblance between the periodic movements in neighbouring glaciers of a range, even if they have the same exposure. Sometimes there is no similarity between the periodic movements in two branches of the same compound glacier; and, occasionally, one side of a glacier tongue may be advancing while the other is stagnant or even retreating.

Case of two glaciers from relatively different local environment: (1) an un-named glacier in the Suru valley of Ladakh, and (2) the Satopanth glacier in Alaknanda valley, Uttarakhand is presented to illustrate the impact of climate change. The first example is from the Trans-Himalayas, an arid environment dependant entirely on winter precipitation with practically nil summer-monsoon precipitation. The second example is from the Central Himalayas with extensive winter and summer-monsoon precipitation. Two glaciers out of 10,000 may not seem to be an adequate representation statistically, but then exception makes the rule. Intensive study programme may bring forth more such examples in support of or disprove the whole concept.

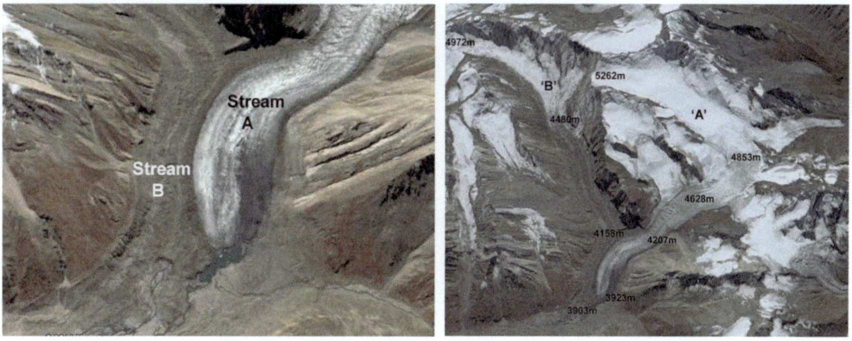

Fig. 2.7 (**a**) Un-named glacier, Suru valley, Ladakh, India, at the end of the ablation season. Two streams (a and b) exhibiting different physical characters. (Google imagery 2007 Highlighted); (**b**) Degenerated snout of the debris covered glacier stream (**b**) continues to extend to a lower altitude than that of debris clear active stream (**a**)

2.3.1 Un-named Glacier in the Suru Valley, Ladakh

In the case of the un-named glacier (Fig. 2.7a), two streams of the glacier originate at different altitudes with a variation of almost 300 m, with accumulation zone of stream 'A' being practically double in width than that of 'B' though orientation is more or less same. Two glacier streams exhibit entirely different character in the ablation zone and along the individual snout. While the stream 'A' shows an active/advancing character; the stream 'B' shows a highly degenerated snout front. It may be noted that stream 'A' has many active feeders while the stream 'B' has practically none.

The main difference between the two streams is the size and orientation of the ablation zone. In the case of stream 'A' it is primarily north-east while that of stream 'B' it is north-west. Also, the gradient of the ablation zone of the stream 'B' is relatively shallow, that of the stream 'A' the gradient is steep with two ice falls. Former shows a high concentration of debris cover and a dirty looking ablation zone, while in the case of latter very thin debris cover is present in the ablation zone. Stream 'B', till 2009, despite the degenerated and debris covered surface of the ablation zone extends slightly to lower altitude than the active stream 'A' (Fig. 2.7b).

2.3.2 Satopanth Glacier, Uttarakhand

Another example selected to put across the point of view under discussion is that of the Satopanth glacier in Uttarakhand. This glacier exists in an entirely different environment than that of the Suru valley. Observations of this glacier and its surroundings, at different times, with the help of pictures taken intermittently over a period of 50 years (1959–2010) have revealed very interesting and intriguing features. As is evident from the comparison of the picture of the year 1959 AD, (Fig. 2.8a) with that of the imagery of 2004 (Fig. 2.8b), no major shift upwards

Satopanth glacier, aerial view, 1959; note the position of snowline

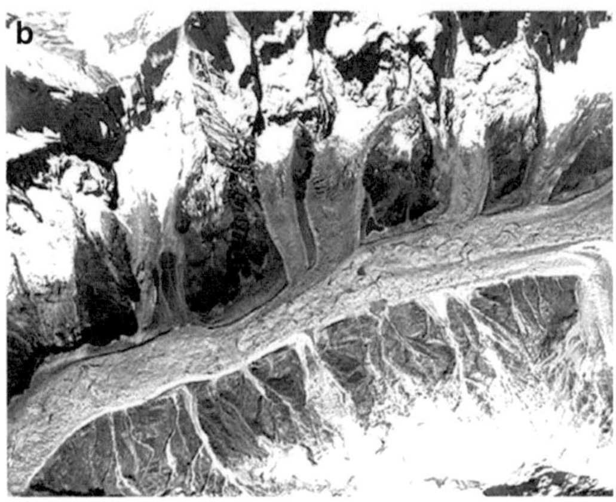

Satopanth glacier, imagery of September 2004; note the position of snowline

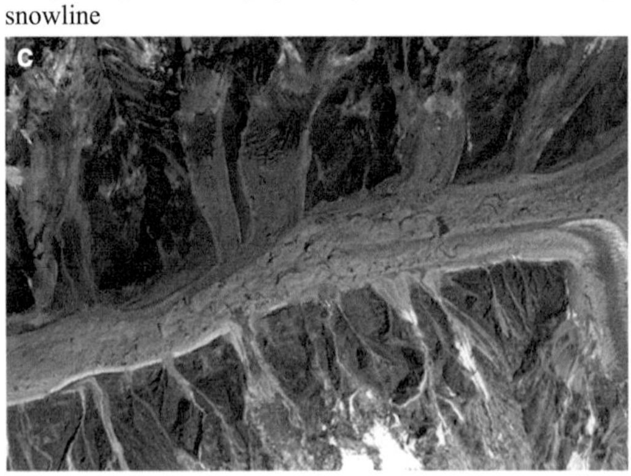

Satopanth glacier, imagery of September 2006; snowline has shifted out ofthe picture frame

Fig. 2.8 Upward shift in the altitude of the snow line (not necessarily reflected at the snout)

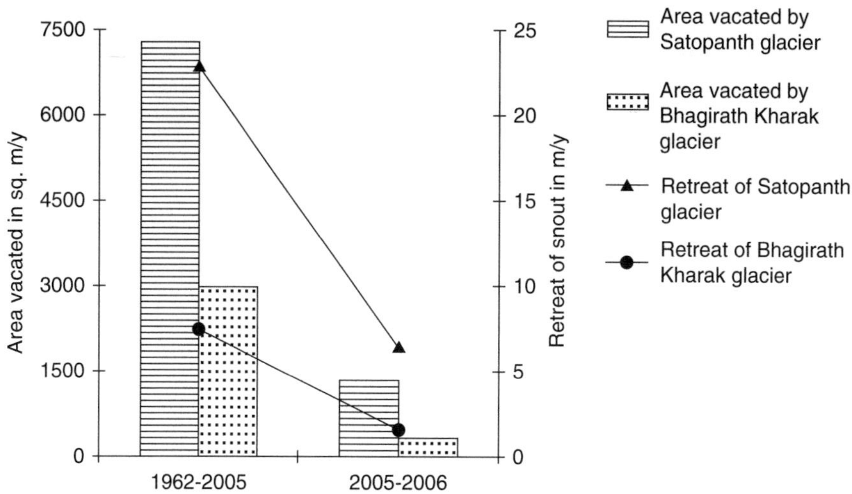

Fig. 2.9 Relative shift (retreat) in the position of the snouts of Satapanth and Bhagirath Kharak glaciers and area vacated from 1962 AD to 2006 AD (Data made available by Prof. Nainwal)

in the snow line is noted for almost 50 years 1959–2004 (presuming it remained so in the intervening period?). During this period, as per the available records, a substantive change (shift upward, almost drastic) has been recorded in the position of the snout of this glacier (Fig. 2.9) with a computed annual average retreat of the order of 25 m.

On the other hand, as can be discerned from the imageries of 2004 and 2006 (Fig. 2.8b, c), there has been a drastic upward shift in the snow line during the 2 year period. Field observations have, however, revealed relative reduction in the annual retreat of the glacier front that came down to less than 5 m; a definite evidence substantiating the basic fact that the glaciers do not necessarily respond to immediate climatic changes.

The examples cited above, un-named glacier in Suru valley, Ladakh and Satopanth glacier in Alaknanda valley, Uttarakhand, are clear evidence to the fact that the glacier snout fluctuation is not influenced by one single parameter but by a combination of parameters. Physiographic character of the accumulation zone and valley slope probably has more dominant role than the annual precipitation and the atmospheric temperature *per se*. For example, the response of a glacier to the meteorological parameters will depend a lot on the slope of the underlying bed rock. A steep glacier is less prone to degeneration as compared to a level glacier, under similar climatic conditions. Possibly transfer of glacier mass from higher levels towards the ablation zone in the case of steeper slopes compensates the greater ablation as compared to low gradient glaciers.

Table 2.2 Temperature data of AWS (SASE)

Year	Bhojwasa temperature (°C)	Kalindikhal temperature (°C)	Nandanban temperature (°C)
2004	5.51	(−)1.35	2.06
2005	2.75	(−)4.80	(−)2.71
2006	3.74	(−)3.70	0.45
2007	4.05	(−)3.80	(−)4.41
2008	2.93	(−)4.25	2.39
2009	1.00	(−)3.52	(−)3.62

2.4 Concluding Remarks

The understanding of the response of the Himalayan glaciers to climatic changes is fraught with a major handicap – there is very little data on the annual snow precipitation and the atmospheric temperature in the Himalaya, in general, and in the fourth order basins that are glacier bearing, in particular, is available. Automatic weather station (AWS) was established at Bhojbasa, about 3 km downstream of the Gangotri glacier snout, in the year 1999, followed by similar establishment at Nandanban and kalindiukhal in the Gangotri glacier area. These AWS's record and relay hourly fluctuations in atmospheric temperature, daily snow/rain precipitation, wind velocity and humidity. Atmospheric temperature in Gangotri glacier basin (see Table 2.2) has shown lowering of the temperature in the last 6 years. Fluctuations recorded along the glacier front may, as in the case of Gangotri glacier for the period 2004–2009, some times show close relation with the climatic parameters but these could be mere coincidence and, it would not be proper to come to any conclusion with the limited data available. One has to remember that whatever studies we make on the glaciers are mainly supra-glacial and we have very little knowledge of what is happening en-glacial or sub-glacial. Further, the phenomenon of global warming and local surface temperature changes, particularly in the glaciated terrain, operate at two very different scales; a direct correspondence may or may not exist. A complete understanding of the response of the Himalayan glaciers to climate change needs a long-term data and an extensive network of monitoring stations.

References

Bhambri R, Bolch T, Chaujar RK, Kulshreshtha SC (2011) Glacier change in Garhwal Himalaya, India, from 1968 to 2006 based on remote sensing. J Glaciol 57(203):543–556

Owen LA, Gualtieri L, Finkel RC, Caffee MW, Benn DI, Sharma MC (2001) Cosmogenic radionuclide dating of glacial landforms in the Lahul Himalaya, northern India: defining the timing of Late Quaternary glaciation. J Quat Sci 16:555–563

Raina VK (2006) Glacier retreat and global warming – a review. XIX Indian Institute of Geomorphologists, National Conference on Himalaya to Indian Ocean, Geomorphic Processes and Landscape Change, University of Jammu, India (in press)
Raina VK, Sangewar CV (2007) Siachen glacier of Karakoram mountains, Ladakh – its secular retreat. J Geol Soc 70:11–16
Raina VK, Srivastava D (2008) Glacier atlas of India. Geol Soc, Bangalore, India
Sangewar C, Shukla SP (2009) Inventory of the Himalayan glaciers. Geol Surv Ind Spl Pub 34:588

Chapter 3
Asynchronous Behavior of Glaciers of Ladakh Himalaya, J & K State, India

R.K. Ganjoo and M.N. Koul

3.1 Introduction

The Himalayan region in India is studded with 9,500 glaciers that vary in size from as small as 1 km in length (unnamed) to as big as 73 km long Siachen glacier. The glaciers of the Himalaya received much attention after the Assessment Report 4 of Intergovernmental Panel on Climate Change (IPCC) made a rather controversial statement that the glaciers in Himalaya shall disappear by 2035. Undoubtedly, the glaciers are considered to be the best indicators to understand the minutest change in climate. This report invited serious attention from the glaciologists, geologists and climatologists of the country and a reassessment of the available data on the glaciers of Himalaya has become necessary. Unfortunately, the inaccessible and treacherous terrain of Himalaya has discouraged many scientists from conducting detailed field-based studies on the glaciers that are located high up and interior of the Himalaya. Despite the fact that Himalaya houses a large number of glaciers, only a few (barely 50 in number) have been studied in detail so far. Nevertheless, the intensive study of a few glaciers has provided sufficient insight that is in contradiction to much hyped suggestion of complete vanishing of glaciers from Himalaya in near future.

R.K. Ganjoo (✉)
Department of Geology, University of Jammu, Jammu, India
e-mail: ganjoork@rediffmail.com

M.N. Koul
Department of Geography, University of Jammu, Jammu, India

R. Sinha and R. Ravindra (eds.), *Earth System Processes and Disaster Management*,
Society of Earth Scientists Series 1, DOI 10.1007/978-3-642-28845-6_3,

3.2 Glaciers of the Ladakh Himalaya

The Ladakh Himalaya houses a total of 4,463 glaciers in its two major basins, namely Indus and Shyok (Sangewar and Shukla 2009). The Indus basin houses a total of 1,796 glaciers and Shyok basin hosts 2,667 glaciers (Sangewar and Shukla 2009). The glaciers in Indus basin are spread over eight sub-basins, namely Katicho, Suru, Leh, Sanglemo Chu, Zanskar, Chhabe Nama, Indus and Hanle. On the other hand, the glaciers of Shyok basin are spread over nine sub-basins, namely Thatte, Saltoro, Sumdo, Fasten, Nubra, Rangdo, Shyok, Chumchar and Chang Chemo. Out of the total area of 43,334 km^2 of the Indus basin, about 2,225 km^2 of area is covered by glaciers, whereas the glaciated area in the Shyok basin is 7,105 km^2 out of the total 37,826 km^2 of the basin (Sangewar and Shukla 2009). Some of the largest glaciers, such as the Siachen (73 km long), glacier identification numbers 5Q131 07 343 (43.75 km long), and 5Q131 02 053 (31.25 km long), are in the Shyok basin (Sangewar and Shukla 2009).

About 53% of the total glaciers in the Indus basin are less than 1 km in length and about 43% of the total glaciers range between 1 and 5 km in length. Similarly, about 45% of the total glaciers in Shyok basin are less than 1 km in length and about 50% of the total glaciers vary between 1 and 5 km in length (Sangewar and Shukla 2009). This reflects the existence of large percentage of small sized glaciers in the Ladakh Himalaya.

The winter of Ladakh is harsh, severely cold and extremely dry. The winter season is from October to May when temperature dips as low as −21°C. The peak winter months are between December and February when mean monthly temperature varies between −14°C and −0.9°C. January is the coldest month. The Western Disturbances cause wide range of fluctuation during winters and also result in major (49%) precipitation in the form of snow in this part of Himalaya. The snowfall in Ladakh Himalaya begins in late October and increases in frequency from December to April.

3.3 Morphology and Dynamics of the Glaciers of the Ladakh Himalaya

Study of some large glaciers representing the Zanskar, Suru and Nubra sub basins of Indus and Shyok basins were undertaken with an objective to unravel the cause for the secular movement of snout. The studies included both field observations and study of IRS satellite data (wherever available) for different time periods (for a time series between 1989 and 2001). The Zanskar sub-basin is represented by Durung Drung glacier, the Suru sub-basin is represented by Kangriz glacier and the Nubra sub-basin is represented by Siachen glacier.

The methodolgy adopted to study the secular movement in the snout of the glaciers is described in the flow chart below:

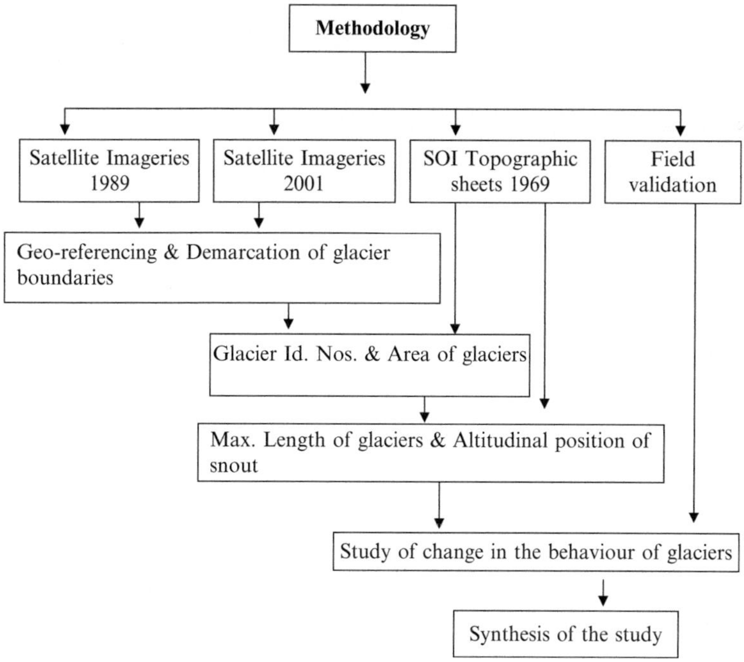

3.3.1 Durung Drung Glacier

The Zanskar basin forms a vast mountainous region between the Great Himalaya Range and Tethyan Himalaya and occupies southern part of the Ladakh Himalaya. The Zanskar basin is studded with 11 major glaciers, namely Durung Drung, Haptal, Haskira, Kange, Hagshu, Sumcha, Lechan, Denya, Mulang, Yaranchu, and Gompi covering an area of 1,098 km^2. The summer snowline (Equilibrium Line Altitude – ELA) rises transversely from SW to NE but is slightly lower on NE than on SW side. Accordingly, the ELA is above 5,800 m on the south but at 5,500 m on the north. The present-day climate of Zanskar valley is arid to semi-arid (Koul et al. 2007).

Durung Drung glacier is 23 km long (Fig. 3.1) and covers an area of 72.83 km^2 (accumulation area 45.68 km^2 and ablation area 27.15 km^2). The glacier, adjoining Pensi La pass that connects the Zanskar with Suru basin, is fed by two major feeders from east and west. The glacier body is situated on Zanskar fault (=South Tibetan Detachment System) (Dèzes 1999) that separates the Himalayan Crystalline Zone from the meta-sediments of Tethyan Himalaya. The reactivation of Zanskar Fault in the younger geological times, late Pleistocene-Holocene, has been responsible for the block movement on the west (Pensi La) compelling the glacier to turn north and further northeast (Ganjoo 2009). This has resulted in a moderate dip of 25° in the

Fig. 3.1 Field view of Durung Drung glacier (August 2007)

Fig. 3.2 Supra-glacier melt water channels on the surface of Durung Drung glacier

east part of the glacier body. The glacier body in the west shows overlapping of layers caused due to the tectonics in the area. The feature thus produced gives a false feeling of surge in glacier.

The glacier surface is criss-crossed both by longitudinal and transverse crevasses. Well developed moulins are on the west flank of the glacier body. The eastern part of the glacier is covered with thin debris. The debris includes large erratic to fine grained sand. Large streams of melt water run on the surface of the glacier. The streams are as wide as 2 m and as deep as 1.5 m (Fig. 3.2). In the

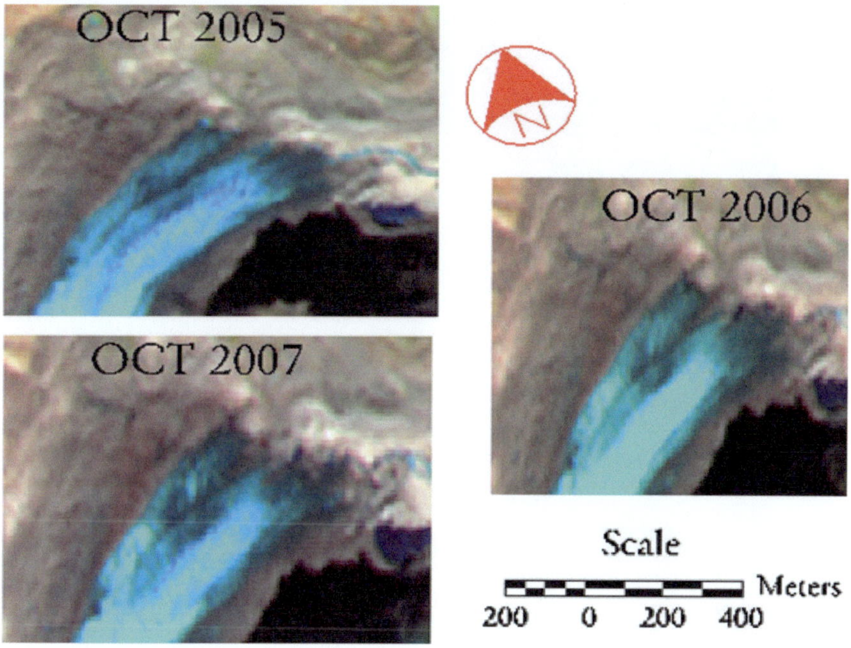

Fig. 3.3 Snout of Durung Drung glacier as seen in IRS imageries of 2005, 2006 and 2007 (peak ablation period)

central part, the melt water channel extends to a length of 500 m extending from an altitude of 4,400 m to snout.

The ablation of glacier is very rapid and maximum in the month of July that varies from 0.81 to 2.7 cm/day. The right (east) flank of the glacier shows very high melting ranging from 0.91 to 3.2 cm/day. The middle part of the glacier has the lowest melting rate 0.81–1.2 cm/day, whereas the left (west) flank of the glacier melts at the rate of 0.87–2.1 cm/day.

The present day snout of Durung Drung glacier at 4,100 m, has caved in the middle and west part resulting in a retreat at an average rate of 3.96 m/year whereas an advance is recorder in the east at an average rate of 0.65 m/year. The degeneration of west part of snout of Durung Drung glacier could be due to several reasons, such as basin geomorphology, location of glacier on active fault, asynchronous behavior of west feeder of the glacier, etc. and not necessarily due to climate change. The study of secular movement of terminus of Durung Drung glacier for past 4 years (2005–2009) thus reveal insignificant change in the snout position that is sufficiently corroborated by field evidences (Ganjoo 2009) and study of satellite imageries (Fig. 3.3).

Fig. 3.4 Snout of the Kangriz glacier in Suru river (August 2007)

3.3.2 Kangriz Glacier

The Kangriz glacier in the Suru valley was undertaken for the study of secular movement of snout and the effect, if any, of inter- and intra-annual seasonal variation on the health of glacier. The 13.48 km long Kangriz glacier emanates from the Kangriz-Shafat ice fields. The snout of the glacier is situated at 3,560 m amsl and debouches in Suru River (Fig. 3.4). The glacier covers an area of 39.94 km^2 out of the total basin area of 88.87 km^2. The glacier is located in a large ice field from which different glaciers, such as Shafat, Techafarka, Branktan, Z and Pinnacle fan out. The accumulation zone of the glacier is situated in the High Himalayan Crystalline of Himalaya that forms a continuation of the Zanskar Ranges.

The Kangriz glacier does not show any change in the position of snout since 1902 (Workman 1909) (Fig. 3.5). However, it is evident from the comparison of the photographs of 1902 and 2007 that the thickness of glacier at snout and ablation zone has certainly reduced in past one century but with no retreat in the position of the snout.

The documentary evidences from travelogue (Neve 1913) when compared with the recent photograph of 2007 (Fig. 3.6), once again, do not show much change in

Fig. 3.5 Note the change in the thickness of the snout of Kangriz glacier (Photograph of 1902 courtesy Sh. V.K. Raina)

Fig. 3.6 Kangriz glacier as documented in 1910 and 2007

the health of glacier. Thus, the documentary evidences for the past one century confirm that the Kangriz glacier has undergone some change in its volume but not in the position of snout. The Kangriz glacier is situated on the Zanskar Fault, similar to that of Durung Drung glacier, and reactivation of fault in geologically younger times (Ganjoo 2009) has resulted in shift of the path of glacier (Fig. 3.7). It would be premature to conclude if the change in volume is exclusively due to change in climate or due to neotectonics in the area.

Fig. 3.7 Note change in the path of movement of Kangriz glacier in geologically younger times

Fig. 3.8 Snout and valley of Siachen glacier – July 2008

3.3.3 Siachen Glacier

The Siachen is the largest glacier, in general, in NW Himalaya and in Karakorum Mountains (Indian part), in particular. Since 1848 the snout of 73 km long valley glacier (Fig. 3.8) has been under observation. The study (Raina 2006), based on available information-report, field photographs and maps, in last about 158 years suggest the following secular retreat/advance of the Siachen glacier snout:

(a) 1862 AD–1909 AD: Advanced by 725 m at an average of 15.42 m/year.
(b) 1909 AD–1929 AD: Retreated by about 50 m at an average rate of 2.50 m/year.
(c) 1929 AD–1958 AD: Retreated by about 400 m at an average of about 14 m/year.
(d) 1958 AD–1985 AD: No change recorded.
(e) 1985 AD–2004 AD: Annual retreat of less than 3 m/year.
(f) 2004 AD–2005 AD: Hardly any retreat has been noted.

The Nubra valley hosts 114 glaciers of various dimensions. Almost 52.6% of the glaciers are less than 5 km long and 31.5% of the total glaciers vary between 5 and 10 km in length in the Nubra valley. This clearly suggests that the Nubra valley is occupied by a large number of small glaciers. A majority of the large glaciers are confined to the southeast segment. The SE facing glaciers comprise nearly 84% of the total glacier area of 114 glaciers. The comparative study of SOI topographic sheets of 1969, Landsat imageries of 1989 and IRS LiSS-III imageries of 2001 suggest beyond doubt that the glaciers of the Nubra valley have not changed much in their length and area between 1989 and 2001 compared to the time period 1969–1989 (Ganjoo et al. 2010).

The study of SOI topographic sheets and satellite imageries for a period of 1969 and 1989 suggests a considerable decline in area from 994.99 km^2 in 1969 to 932.90 km^2 in 1989. However, no significant change in the area of Siachen glacier is noticed for a period of next one decade from 1989 to 2001. The satellite imageries of 1989 and 2001 suggest a decline in the area from 932.90 km^2 in 1989 to 930.37 km^2 in 2001. This is also reflected in the change in maximum length of the Siachen glacier from 68.30 km in 1969 to 67.56 km in 1989 to 67.50 km in 2001. The negligible vacation of area and reduction in the maximum length of the Siachen glacier, particularly during the period between 1989 and 2001 is further corroborated by the field studies made in July 2008 (Ganjoo and Koul 2009).

Asynchronous behavior is not only noticed in the large glaciers, such as Siachen, but also in small sized glaciers. A small sized glacier identified as Glacier no. 11 of Nubra valley is independent and lies northeast of Siachen glacier. The comparative study of imagery of 1989 and IRS LiSS-III imagery of 2001 (Fig. 3.9) clearly shows an increase in the length and area of the glacier over a period of three decades. The said glacier has a maximum length of 2.9 km and area of 0.98 km^2 in 1969 that increased to 3.5 km and 1.12 km^2, respectively in 1989. The IRS LiSS-III imagery of the glacier of 2001 shows a minor decrease in the length of glacier to 3.2 km and of area to 1.07 km^2. The said glacier, therefore, has certainly undergone change since 1969 that may not necessarily be attributed to change in climate.

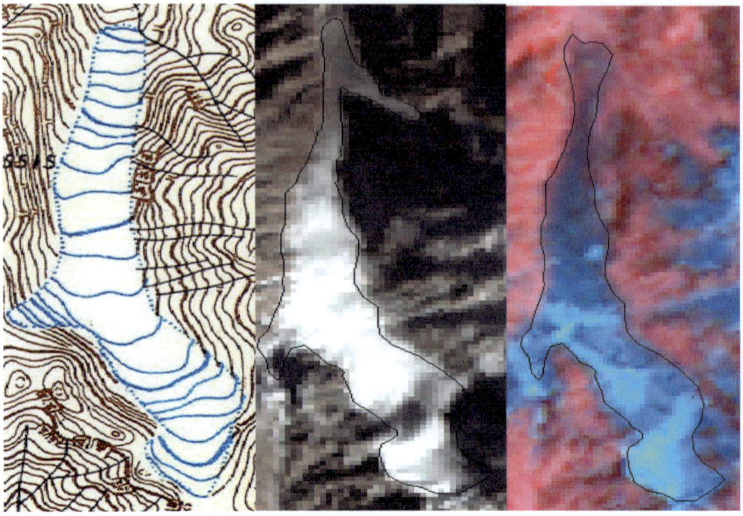

Fig. 3.9 Glacier no. 11 of Nubra valley

3.4 Discussion

In general it has been noticed that the Himalayan glaciers show decline in their rate of recession. The recession of Gangotri glacier (Garhwal Himalaya) has reduced to 12 m/year in 2004–2005 (Kumar et al. 2008) from 26 m/year between 1935 and 1971 (Raina 2003; Sharma and Owen 1996; Naithani et al. 2001; Srivastava 2003). The same is true for Pindari glacier (Kumaon Himalaya) where the rate of recession has been calculated to 6.5 m/year between 1966 and 2007 (Bali et al. 2009) compared to 26 m/year between 1845 and 1906 (Tewari 1973). In case of Dokriani glacier the rate of recession has been more or less constant (16–18 m/year) throughout the period between 1962 and 1995 (Dobhal et al. 2004). The Satopanth glacier that was receding at the rate of 22.8 m/year has now reduced to 6.5 m/year during 2005–2006 (Nainwal et al. 2008).

The Ladakh Himalaya holds special geographical significance to study the effect of climate change on the glaciers. It is because of the fact that Ladakh Himalaya is situated far off from the effect of South West Indian Monsoon (SWIM). The SWIM is considered to be a major moisture carrier to the Himalaya and the moisture rich winds nourish the glaciers situated in most of NW Himalaya, e.g. Uttarakhand, Himachal Pradesh. The Ladakh Himalayas, on the other hand, receive most of the precipitation through Westerlies during the winters.

The study of three major glaciers in Ladakh Himalaya has provided adequate insight to the complex behaviour of the glaciers. The Siachen glacier, throughout its monitoring for past more than a century and half has shown the erratic behaviour in the secular movement of the snout. An extreme advance of 725 m between 1862 and 1909 and quantum retreat of 400 m between 1929 and 1958 in the snout position of the Siachen glacier is beyond any scientific explanation. The Siachen glacier has completely slowed down in its retreat for almost a decade which again is not explainable by any scientific logic. Similarly, the Durung Drung glacier has not shown much deterioration in the snout position of the glacier in past about 6 years. Interestingly, the east and west feeders of the glacier behave asynchronously to the similar geological, meteorological and geographical conditions. The asynchronous behaviour between the feeders of the Durung Drung glacier has partial impact on the change in the snout position of the glacier. The Kangriz glacier has not shown any change in the position of the snout for past one century. However, the thickness of the glaciers has reduced since 1907 causing change in the volume of the glacier.

Data presented in this paper clearly shows that the changes in the glaciers of Ladakh Himalaya are varied and complex. Some glaciers show change in length with respect to area, such as the Kangriz glacier, whereas some glaciers show variation in area and not in length, such as the Siachen glacier. The health of glaciers in Ladakh Himalaya are affected by several factors, such as inter- and intra-annual variation in weather parameters (micro-climatic factors), terrain morphology, aspect of glaciers, neotectonics, etc. Studies on the effect of weather parameters on the health of Naradu glacier (H.P.) between 1994 and 2004 clearly suggest that relationship between secular movement of snout and weather parameters is complex and varied (Koul and Ganjoo 2010).

3.5 Concluding Remarks

In the absence of intensive, long term meteorological data from the high mountains of the Ladakh Himalaya, and a sound modeling for predicting the change, it would be premature to attribute the secular movement of snouts exclusively to climate change. However, it is beyond doubt that glaciers of the Ladakh Himalaya do not show significant change in their length and area over a span of past one century.

Acknowledgements The substantial funding received in the form of various research projects from Department of Science and Technology, New Delhi and Space Applications Centre, Indian Space Research Organisation, Ahmedabad to conduct the research studies in Ladakh is duly acknowledged.

References

Bali R, Agarwal KK, Ali SN, Rastogi SK, Krishna K (2009) Monitoring recessional pattern of Central Himalayan glaciers: some optimistic observations. Proc Indian Sci Congr 96:79–80

Dèzes P (1999) Tectonic and metamorphic evolution of the Central Himalayan domain in southeast Zanskar (Kashmir, India). Mémoir Géol (Lausanne) 32.1–149

Dobhal DP, Gergan JT, Thayyen RJ (2004) Recession and morpho-geometrical changes of Dokriani glacier (1962–1995) Garhwal Himalaya, India. Curr Sci 86(5):692–696
Ganjoo RK (2009) Holocene tectonics and climate of Durung Drung glacier basin, Zanskar Himalaya, India (Abstract).The fifth international symposium on Tibetan plateau and 24th Himalaya-Karakorum-Tibet workshop, Beijing, China, 11–14 Aug 2009
Ganjoo RK, Koul MN (2009) Is the Siachen glacier melting? Curr Sci 97(3):309–310
Ganjoo RK, Koul MN, Ajai, Kulkarni AV, Bahuguna IM (2010) Snow and glacier studies of Nubra valley including Siachen. Project Completion Report (unpublished). Space Applications Centre, Ahmedabad
Koul MN, Ganjoo RK (2010) Impact of inter- and intra-annual variation in weather parameters on mass balance and equilibrium line altitude of Naradu glacier (Himachal Pradesh), NW Himalaya, India. Clim Chang 99:119–139
Koul MN, Ganjoo RK, Dhinwa PS, Pathan SK, Ajai (2007) Use of geo-informatics for preparation of desertification status map of Stod valley (1F4C2), Padam (Zanskar), District Kargil, J & K State. Indian Soc Geomat Newsl 13(2–3):27–35
Kumar K, Dumka RK, Miral MS, Satyal GS, Pant M (2008) Estimation of retreat of Gangotri glacier using rapid static and kinematic GPS survey. Curr Sci 94(2):258–262
Nainwal HC, Negi BDS, Chaudhary M, Sajwan KS, Gaurav A (2008) Temporal changes in rate of recession: evidences from Satopanth and Bhagirath Kharak glaciers, Uttarakhand, using total station survey. Curr Sci 97(5):653–660
Naithani AK, Nainwal HC, Prasad CP (2001) Geomorphological evidences of retreat of Gangotri glacier and its characteristics. Curr Sci 80:87–94
Neve A (1913) Thirty years in Kashmir. Edward & Arnold, London
Raina VK (2003) History of Gangotri glacier down the ages. Proceeding of workshop on Gangotri Glacier 2003. Geological Survey of India Special Publication, vol 80, pp 1–10
Raina VK (2006) Glacier retreat and global warming – a review. Keynote address delivered at XIX IGI conference on Himalaya to Indian Ocean – Geomorphic processes and landscape change, Institute of Himalayan Glaciology, University of Jammu, Jammu, 26–28 Dec 2006
Sangewar CV, Shukla SP (2009) Inventory of the Himalayan glaciers. Geological Survey of India, Calcutta, Special Publication No. 34
Sharma MC, Owen LA (1996) Quaternary glacial history of the Garhwal Himalaya, India. Quat Sci Rev 15:350–365
Srivastava D (2003) Recession of Gangotri glacier. Proceeding of workshop on Gangotri glacier 2003. Geological Survey of India Special Publication, vol 80, pp 21–30
Tewari AP (1973) Recent changes in the snout of Pindari glacier (Kumaon Himalaya): the role of snow and ice in hydrology. Proceedings of Banff symposium, UNESCO-WHO-IAHS, vol 2, pp 1144–1149
Workman FB (1909) Peaks and glaciers of Nun Kun: a record of pioneer exploration and mountaineering in the Punjab Himalaya. Constable & Company Ltd., London

Chapter 4
Signatures and Evidences of Surging Glaciers in the Shyok Valley, Karakoram Himalaya, Ladakh Region, Jammu & Kashmir State, India

Anjani Kumar Tangri, Ram Chandra, and S.K.S. Yadav

4.1 Introduction

The Himalayas are the storehouse of snow, ice and glaciers covering an area of approximately 33,200 km^2 (Flint 1971). Recent study using high-resolution satellite data has revealed that there are 32,392 glaciers in Indus, Ganga and Brahmaputra basins of Himalaya with a total glaciated area of 71,182 km^2. The Indian Himalaya alone has 16,627 glaciers which covers an area of 40,463 km^2 (SAC-ISRO Technical Report 2010). The dimensions of these glaciers are constantly changing in time and space with changing climate. The Himalayan glaciers are an important component of the global cryosphere, which occupy the highest mountain range in the world. As the geological record of the earth indicates, there have been four or five major glacial periods separated by interglacial periods. During that period climate was warmer and deglaciation occurred on large scales. During the peak glaciation in Pleistocene, about 46 million km^2 area was covered by the glaciers which is more than three times the present ice cover of the earth. The Himalayas has one of the largest concentrations of glaciers and are highly sensitive to minor changes in the atmospheric temperature and are also considered as very good indicators of the changing climatic scenarios of the past. The Himalayan glaciers have lately been on a constant retreat and diminishing both in their spatial extent as also in their volume. However, a couple of glaciers specially in the Karakoram Himalaya have demonstrated an altogether new dimension in their behavior and these "surging glaciers" have been monitored.

A.K. Tangri (✉) • R. Chandra • S.K.S. Yadav
Remote Sensing Applications Centre, Lucknow, UP 226021, India
e-mail: aktangri@rediffmail.com

R. Sinha and R. Ravindra (eds.), *Earth System Processes and Disaster Management*, Society of Earth Scientists Series 1, DOI 10.1007/978-3-642-28845-6_4,

4.2 Surging Glacier

Glacial surges are short-lived events where a glacier can move up to velocities 100 times faster than normal, and advance substantially. Surging glaciers are clustered around a few areas across the world. High concentration of Surging glaciers can be found in Svalbard, (Hambrey and Dowdeswell 1997) Canadian Arctic islands, Alaska and a few places in the Himalaya. Glacial surges take place at regular, periodic intervals; the period between two surges is called the "quiescent phase". During this period, the velocities of the Glacier are significantly lower, and the glaciers can retreat substantially. Glacier Surging, popularly known as "galloping", is one of the most dramatic phenomena of glacier motion (Harrison and Post 2003). A Surging glacier, after flowing along an apparently normal manner for years, speeds up for a relatively short time to flow rates as much as 100 times than the normal rates and then drop back to an apparently slower flow rate (Kamb 1985).

Usually Surging events last for less than 1 year and occur periodically, between 15 and 100 years. However, the periodic surges often occur at regular intervals typically between 10 and 100 years. Some of the glaciers exhibit major periodic fluctuations in velocity over time scales ranging from a few years to several centuries, swinging between phases of rapid and slow flow.

Surges can be serious hazards in populated areas, engulfing occupied land, generating sudden floods, or disrupting local communications below and across glaciers. They tend to recur in cycles, but seem unrelated to normal patterns of advance and retreat. In regions with many Surging glaciers, such as the Karakoram Himalaya, surges complicate the normally sensitive relationship between glaciers and climate.

4.3 Location of the Study Area

Karakoram lies immediately north of the western part of the Greater Himalaya and is the highest of the southwest Central Asian mountain systems. It has the largest concentration of glaciers in mainland Asia and outside high latitudes, with 8 Glaciers over 50 km long and more than 20 glaciers more than 30 km long. The perennial snow and ice cover comprises a huge fresh water resource and glacier melt waters make a major contribution to the flow of the Indus and Yarkand Rivers and to the livelihood of some 130 million people. In the last 100 years, 26 sudden, rapid advances involving 17 glaciers have been reported in the Karakoram region. Another 12 glaciers have features associated with surge behaviour.

The northern face of Karakoram Range between the Nubra valley (Siachen glacier) in the west and Aksai Chin in the east is traversed by the Shyok River originating from the Rimo group of glaciers (Table 4.1). The Shyok River passes

Table 4.1 Characteristics of the Kumdan and Rimo group of glaciers

S. no.	Glacier characteristics	Kumdan group of glacier			Rimo group of glacier
		Chong	Kichik	Aqtasti	
1	Total watershed area (km^2)	172,02	85.05	37.54	–
2	Total glacier area (km^2)	105.35	46.36	20.16	–
3	Glacier length (km)	26	18	12	40
4	Glacier width (km)	2.5	1.5	< 1.0	>3.0
5	Glacier orientation	SE	E	E	SE
6	Elevation range (meters a. msl.)	4,700–7,000	4,600–7,000	4,500–6,800	5,100–8,000

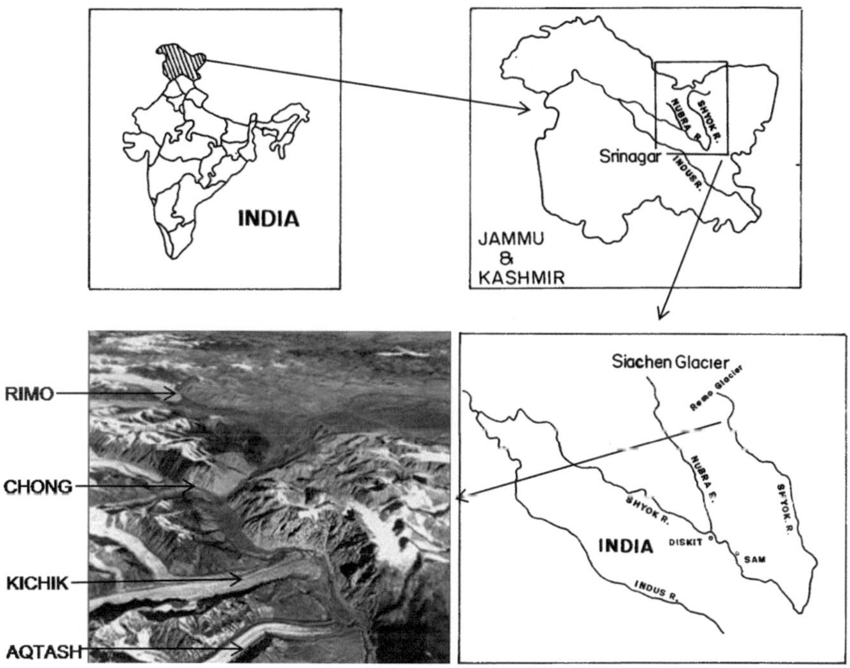

Fig. 4.1 Location of the study area

along the five transverse glaciers i.e. Chong Kumdan, Kichik Kumdan, Kichik (little), Aqtash and Chatrik glacier. The Rimo along with the Chong Kumdan, the Kichik Kumdan and the Aqtash glaciers have been reported to exhibit surging phenomenon (Raina and Srivastava 2008; Tangri et al. 2008, 2011; Ram Chandra and Tangri 2011). The last three glaciers sometime cross the Shyok valley and come in contact with the opposite valley wall blocking the river course and forming huge lakes in the upstream (Fig. 4.1).

4.4 Physiography of the Glaciers Studied

The Rimo glacier is a complex basin glacier having two distinct limbs – the Northern and Southern limb. This 40 km long glacier trending southeast has a width of approximately 3 km. The elevation ranges from 5,100 m at the snout to approximately 8,000 m in the accumulation zone. The Kumdan group of glacier (Table 4.1) comprises of three glaciers namely, Chong Kumdan, Kichik Kumdan and Aqtash.

The Chong Kumdan glacier watershed has an area of 172 km^2 out of which the glaciers occupying 105 km^2.This is also a SE trending glacier and is 26 km long and approximately 2.5 km wide. The glacier takes its name from the Turkish word Chong (Large) and Kumdan (Black kiln) probably because of the dark colour of glacier ice. "Chong Kumdan" also means big obstruction and the name has probably been given because of the huge glacier dam lakes being produced upstream of the glacier due to its surging phenomena. This glacier has three prominent limbs *i.e.* Northern, Central and southern limb.

The Kichik Kumdan glacier watershed has an area of 85 km^2 and the area occupied by the glacier is approx 46 km^2. This is an east trending glacier, 18 km long and approx.1.5 km wide. The elevation ranges from 4,600 m at the snout to 7,000 m in the upper reaches. The glacier takes its name from "Kichik" (Small) and "Kumdan" (Kiln). It is also known as the Thangman glacier. "Kichik Kumdan" also means small obstructions because of the fact that after an ice surge, a small lake is formed behind the obstruction. This glacier also has two limbs i.e. the Northern limb and the Southern limb.

The Aqtash glacier watershed is approx 37 km^2 in area out of which the glacier occupies 20 km^2. This is an east trending glacier, 12 km long and less than 1 km in width. The elevation ranges from 4,500 to 6,800 m. "Aqtash" means white and it also takes its name from the exposure of marble close to it's snout along the western bank.

4.5 Signatures of Surging in the Studied Glaciers

High resolution Quickbird satellite data coupled with the precision geocoded data from Indian Remote Sensing satellite have been visually interpreted and digitally analyzed to monitor different surface features, glacial geomorphological characteristics and other glacial parameters of the four glaciers undertaken in the present exercise.

The Rimo group of glaciers comprise of the northern and the southern limbs. The northern limb has a lobate front with concentric crevasses in the frontal and central part of the glacial body. The two lateral glaciers of small dimensions meeting the trunk of the northern limb from the south are squeezed because of the push from the main body. The trunk glacier shows upward bulge demonstrating the push from

Fig. 4.2 Evidences of surge in the two limbs of Rimo glacier – lobate front, tightly folded glacial surface, zig zag pattern of medial moraine, squeezing of the lateral glaciers by the trunk glacier, over-spilling of glacial ice on to the valley walls

behind. The glacial ice seems to be over spilling the valley limits and climbing on to the valley walls. The southern limb of the Rimo glacier presents an undulating topography so characteristics of the surging glaciers. The surface of the southern limb appears to be tightly folded. Additionally, the lateral glaciers meeting the trunk of the southern limb are all squeezed because of the push of the main body. The medial moraines in the southern limb of Rimo glaciers present an irregular Zig-Zag pattern – a characteristic of the surging phenomenon (Fig. 4.2). The area ahead of the frontal part of the Rimo glacier appears highly excavated, perhaps because of the rapid advance and retreat of the glacier in the past. Drastically different behaviours have been observed in the present day northern and southern limbs of Rimo glaciers. While the northern limb with a lobate front presents the advancing stage, the southern limb is apparently on the retreat.

The Chong Kumdan glacier consists of three limbs – the northern, central and the southern. The northern and the central limbs merge together to form a single lobate front. The medial moraine is being pushed towards the north, demonstrating high rate of advance of the central limb. Both the northern and the central limbs have a lobate frontal part. The southern limb of Chong Kumdan glacier represents a highly eroded and excavated frontal part with clear evidences of the glacier retreating back (Fig. 4.3). The excavated region ahead of the frontal part demonstrates the extent to which the glacier must have advanced during it's earlier cycles of surge. The concurrent advancing stage of the northern and central limbs and the reversed trend of retreat noticed in the southern limb of Chong Kumdan glacier are typical characteristics of the surging phenomenon.

The Kichik Kumdan glacier is a single limb east–west trending glacier which presently has a huge lobate front almost pushing the Shyok river to the opposite river valley wall (Fig. 4.4). This is a definite indication of the advancing stage of this surging glacier.

The Aqtash glacier has an excavated frontal part representing the maximum extent of advance the glacier must have experienced during its earlier surge,

Fig. 4.3 Lobate front, medial moraine being pushes aside, highly excavated frontal area and simultaneous surge and retreat noticed in the adjoining limbs of the Chong Kumdan glacier are all evidences of surging glaciers

Fig. 4.4 Lobate front of Kichik Kumdan glacier, almost pushing the Shyok river to the valley wall and squeezing of the lateral glaciers. Excavated frontal part of the Aqfash glacier on the *left* shows degeneration of the glacial body

wherein it was almost "debouching" against the Shyok river. Presently, the Aqtash glacier seems to be a couple of meters behind it's previous maximum extension.

4.6 Temporal Monitoring of the Surging Phenomenon

Sporadic data on the Rimo, Chong Kumdan, Kichik Kumdan and Aqtash glaciers in Shyok valley of Karakoram Himalaya does exists for almost a century and a half. The earlier observations collected by Central Asian Traders following the silk route, although not based on any scientific footing; still provide an insight into the behavioral pattern of these glaciers from time to time. Authentic record of the studies carried out by the Survey of India and Geological Survey of India date back to almost a century.

High resolution precision geocoded data products from the Indian Remote Sensing satellite in conjunction with the Survey of India topographic sheets has enabled monitoring the recent temporal variations in the overall configuration of these four glaciers in Shyok valley. Survey of India topographical sheets of the study area date back to 1962/1975 and provide an authentic record of the then position of the frontal parts of these glaciers and their overall spatial extents. IRS-1A/1B/1C/1D and P6 satellites have provided the annual scenario of the study area from 1990 to 2009. These satellite data products have been visually interpreted and digitally analyzed. A simple superimposition of these glacier demarcations of different years, in a GIS environment, has enabled precise monitoring of the variations in their frontal parts and their overall configurations. These have been presented herein.

4.6.1 Kumdan Group of Glaciers

4.6.1.1 Chong Kumdan Glacier

The behaviour of the three limbs of the Chong Kumdan glacier have been monitored using the satellite data for the period 1990–2009, in conjunction with an old record of 1975 derived from SOI topographical sheets. The 1975 topographical sheets show that the northern and the central limbs merge into each other and their common front merges into the southern limb. Accordingly, the three limbs meet each other to form a single frontal part almost touching the Shyok River.

Each of the three limbs of Chong Kumdan glacier has behaved in a different manner. The northern limb shows almost 200 m of advance between 1992 and 1996 and then a retreat by almost 250 m between 1996 and 1997. Then after, except for the years 2001 and 2003, the northern limb is continuously advancing and overall advance between 1997 and 2009 has been to the tune of 2,278 m downstream (Figs. 4.5 and 4.6).The central limb also has a history of oscillations. The central limb retreated during 1996, 1997, 2000 and 2001 but in general it shows an overall advance of almost 1,275 m between 1997 and 2009. The southern limb shows a cyclic behavior. Between 1975 and 1990 this limb has retreated by almost 1,250 m. Between 1999 and 2000 this limb advanced by 370 m. Subsequently, between 2000

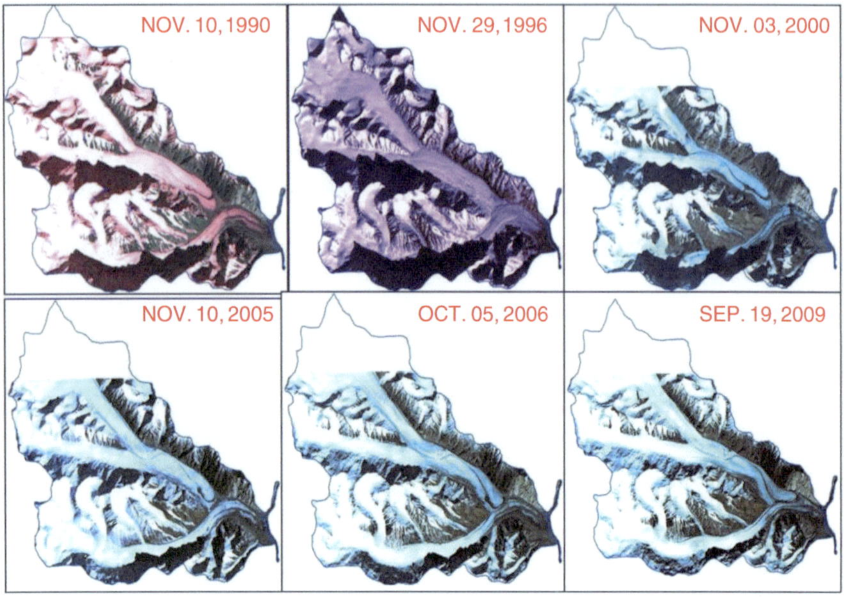

Fig. 4.5 Temporal monitoring of Chong Kumdan glacier using multi-date high resolution IRS satellite data

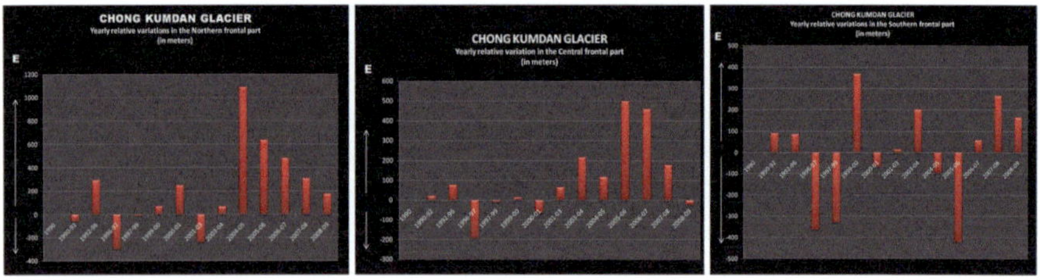

Fig. 4.6 Graphical representation of temporal variations in the frontal part of the three limbs of Chong Kumdan glacier

and 2006 there have been slight oscillations but the overall retreat has been 96 m. Multi-date satellite data demonstrates that the southern limb of the Chong Kumdan glacier has advanced by almost 400 m between 2006 and 2009.

The surging phenomenon is prominently demonstrated by the disposition of medial moraine between the northern and central limb of the Chong Kumdan glacier. Due to the variation in the pressure exerted by the northern or the central limb, the configuration of the medial moraine between the two keeps on changing from year to year.

The excavated frontal part of the southern limb of the Chong Kumdan glacier clearly demonstrates the maximum extent of advance this limb must have experienced in the past. Detailed mapping of the excavated area has helped in monitoring the different stages of retreat of the southern limb. During the stage of its maximum advance, the southern limb must have blocked the Shyok River, thereby forming a large lake upstream.

The present lobate frontal part of the northern, central and also the southern limbs of the Chong Kumdan glacier clearly demonstrate that the totality of the glacier is presently advancing and nearly touching the Shyok river.

4.6.1.2 Kichik Kumdan Glacier

This is an east–west trending single limb glacier with a broad lobate front typical of the surging phenomenon. The first view of this glacier on the satellite data indicates that the glacial ice flowing in the confined valley tends to spread out the moment it reaches the wide flood plain of Shyok River. The wide spreading of the frontal part of the Kichik Kumdan glacier gives an impression that the multi-million tonnes of ice has "flowed" almost to the tune of kilometers in the last few years, behaving more like a fluid than a solid.

Multi-date satellite data in conjunction with SOI topographic sheets have demonstrated that the frontal part of Kichik Kumdan glacier almost pushed the Shyok river to the opposite valley wall in 1975 but it had retreated by almost 750 m by the year 1990. However, with slight reversals during the period 1993–1996, the retreating tendency continued right up to 1996. Thereafter, there was a sudden surge in the frontal part, which advanced by almost 1,200 m between 1997 and 2000. The entire frontal part is again on retreat except for a sporadic surge between 2003 and 2004 (Figs. 4.7 and 4.8).

The maximum advance recorded in the Kichik glacier demonstrates that the glacial front has almost pushed the Shyok river to the opposite valley wall and a short period of blockade of the river and formation of lake in upstream must have existed sometimes in the past. The ponding of the water must not have lasted long as demonstrated by the signature of the lake dimensions on the satellite image. This lake burst either due to the overtopping of the glacier frontal ice by the lake water or the subsequent retreat must have led to enormous devastation downstream.

4.6.1.3 Aqtash Glacier

This is also an east–west trending single limb glacier which has had a continuous history of advance and retreat. Multi-date satellite data demonstrate that the advances in the frontal part have been of higher magnitude when compared to the retreats. These have been alternating over a periodicity of almost 3–4 years except for the present stage where in the glacier is on a continuous advance since 2006. The maximum extent of the frontal part must have "debouched" against the Shyok River in the past. Although of a much smaller dimensions than the other three glaciers

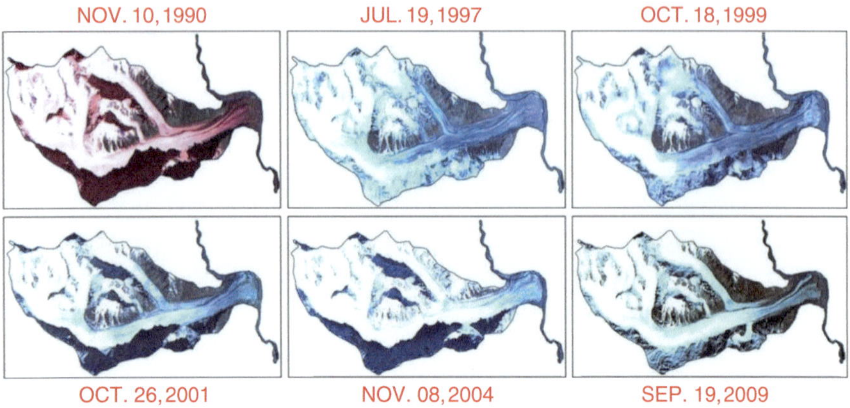

Fig. 4.7 Temporal monitoring of Kichik Kumdan glacier using multi-date IRS satellite data

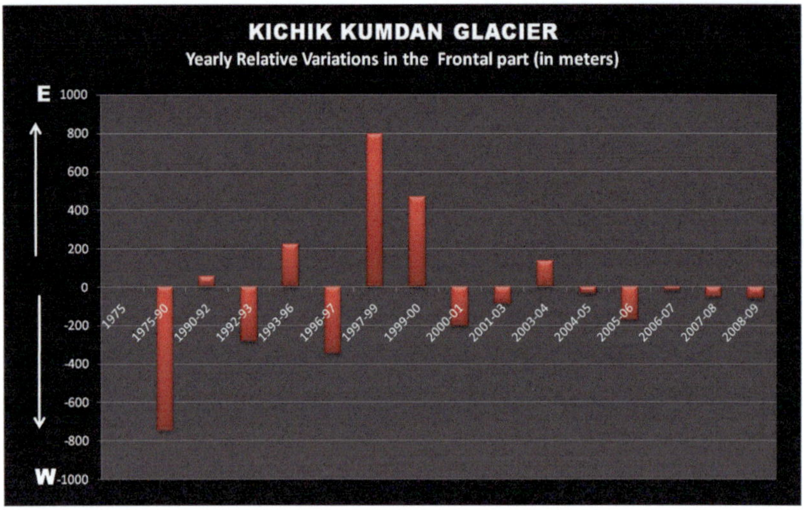

Fig. 4.8 Graphical representation of temporal variations in the frontal part of the Kichik Kumdan glacier

under study, the Aqtash glacier nevertheless demonstrates cyclicity in its overall history of advance and retreat (Figs. 4.9 and 4.10).

4.6.2 Rimo Group of Glaciers

This is a complex basin glacier having two distinct limbs in the Indian territory. Each of these limbs behaves in a different manner but both demonstrate the phenomenon of surging in recent times. The SOI topographical sheet of 1962 has

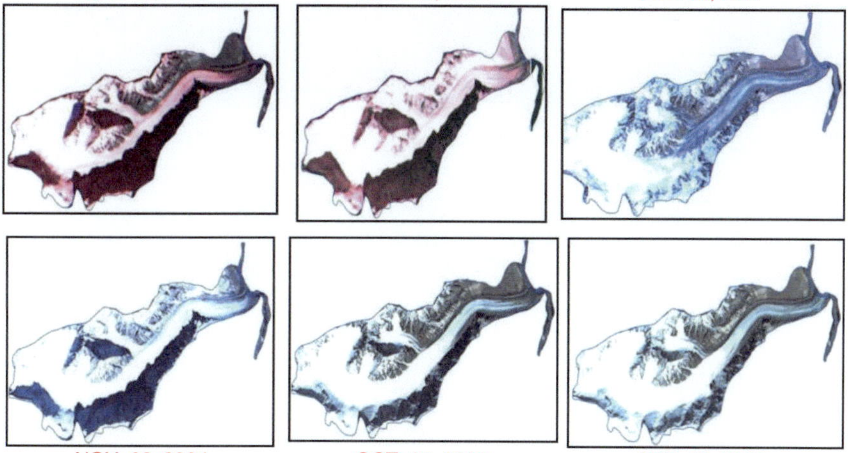

Fig. 4.9 Temporal monitoring of Aqtash glacier using multi-date IRS satellite data

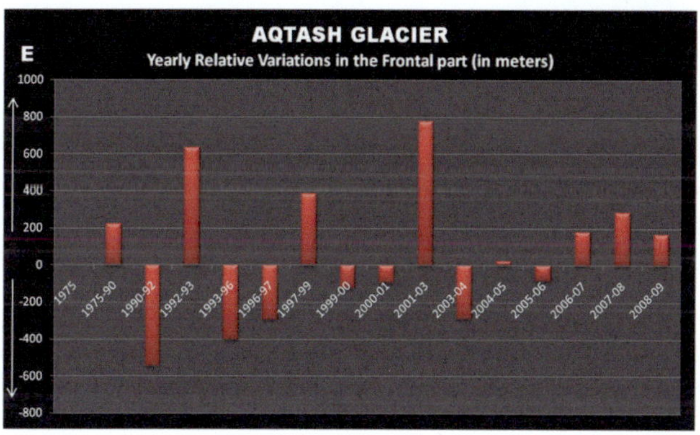

Fig. 4.10 Graphical representation of temporal variations in the frontal part of Aqtash glacier

been taken as the reference and successive variations have accordingly been monitored using multi-date satellite data.

With respect to 1962 position, the 1990 satellite data demonstrates that during this period, there has been a retreat of more than 1 km in the northern limb. Except for reversal during 1995–1996, the northern limb experienced a continuous advance up to 1997 to the tune of almost 1.8 km. The frontal part of the northern limb has remained almost in "quiescent stage" since then with a broad lobate frontal part (Fig. 4.11).

The southern limb of the Rimo glacier shows an advance of ~125 m between 1962 and 1990 and a further advance of almost 150 m during 1994–1995. Then after

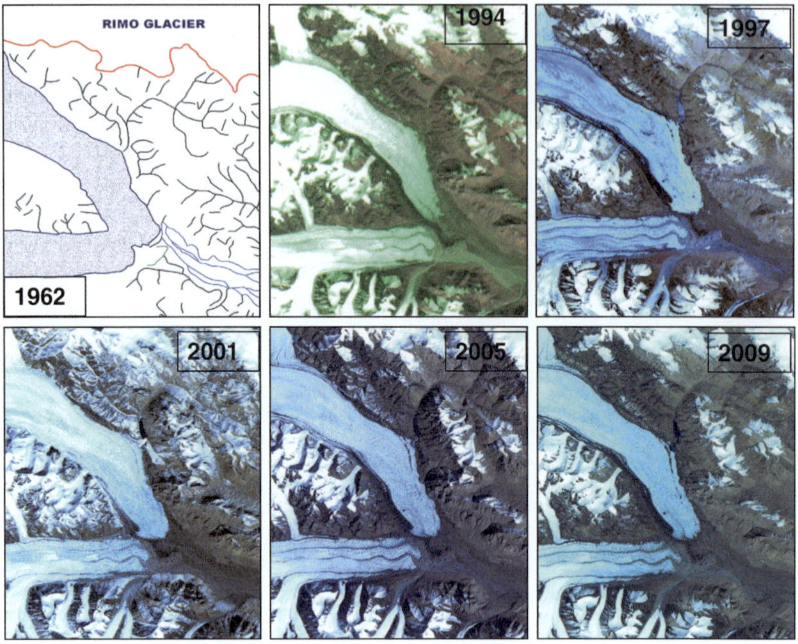

Fig. 4.11 Temporal monitoring of Rimo glacier using multi-date IRS satellite data

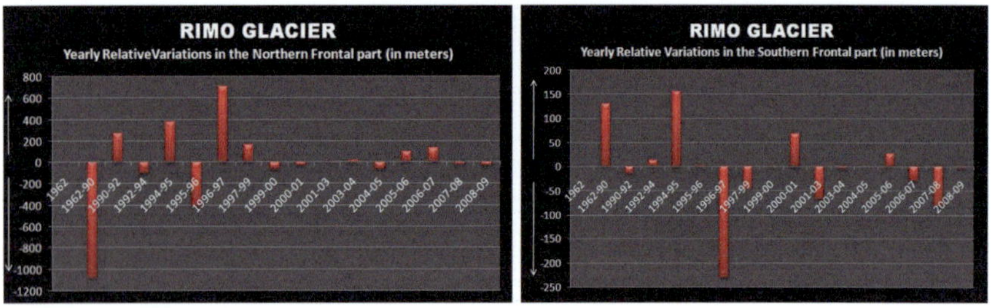

Fig. 4.12 Graphical representation of temporal variations in the frontal part of Rimo glacier

this, southern limb has experienced a continuous retreat which continues even today (except for small reversals in the year 2000 and 2005) (Fig. 4.12).

4.7 Concluding Remarks

Historical records indicate that the four glaciers under study i.e. the Rimo, the Chong Kumdan, the Kichik Kumdan and the Aqtash have had a history of surge and some sort of cyclicity in their behavior (Tables 4.2 and 4.3). Each of these glaciers has been from time to time, either advancing or retreating or being in a quiescent

Table 4.2 Temporal variations in the frontal part of the Rimo group of glaciers

Name	Limb	Period (in years)	Distance shift (in meters)	
Rimo group of glacier	Northern	1962–1990	1,083	←
		1990–1995	377	→
		1995–1997	708	→
		1997–2001	34	←
		2001–2008	19	←
	Southern	1962–1990	129	→
		1990–1995	155	→
		1995–1997	230	←
		1997–2001	67	→
		2001–2008	82	←

(Retreat ←) (Surge →)

Table 4.3 Temporal variations in the frontal part of the Kumdan group of glaciers

Name		Limb	Period (in years)	Distance shift (in meters)	
Kumdan group of glacier	Chong Kumdan glacier	Northern	1990–1992	50	←
			1992–1996	300	→
			1996–1997	300	←
			2004–2005	1,100	→
			1997–2009	2,278	→
		Central	1997–2005	355	→
			2005–2007	955	→
			2007–2009	23	←
		Southern	1975 1990	1,250	←
			1999–2000	370	→
			2000–2006	96	←
			2006–2009	400	→
	Kichik Kumdan glacier	Single lobe	1975–1990	750	←
			1990–1996	33	←
			1996–2000	900	→
			2000–2009	550	←
	Aqtash glacier	Single	1975–1992	300	←
			1992–2003	830	→
			2003–2009	192	→

(Retreat ←) (Surge →)

stage. Even the individual limbs of each of these glaciers have shown characteristic behaviour, distinctly apart from the adjoining limb. At times, the totality of the glacier behaves in a similar fashion, while at other times they behave differently.

Since the Shyok valley is a confined valley on the northern slopes of the Karakoram Himalaya, it could be presumed that the overall climatic variations in the valley would be more or less of the same order. Moreover, the frontal parts of these glaciers are also in the same elevation zones and hence altitudinal variation is not much. As such the causative factors for the distinctly different advancing or retreating behavioral pattern of individual limbs of the glacier is still not well understood (Barrand and Murray 2006; Wake and Searle 1993).

The sudden advance of these glaciers, at times to the magnitude of a couple of kilometers over span of few years, is an amazing factor. This could many a times, result in blocking the course of Shyok river forming huge glacial dammed lakes in the upstream. The Shyok valley is quite conspicuous by the existence of broad floodplains upstream of the Chong Kumdan, the Kichik Kumdan and the Aqtash glaciers. These are perhaps caused by the lake fill deposits during course of blocked of the Shyok River by these surging glaciers (Hewitt 1969, 2007). However, a great deal of effort and investigations are required to be able to comprehend the reasons for this exceptional glacial phenomenon.

Acknowledgements Authors are indebted to the Director, RSAC-UP, Lucknow for kind permission to undertake these investigations and for providing necessary working facilities. Financial support from the Department of Science & Technology, Govt. of India, New Delhi is also gratefully acknowledged.

References

Barrand NE, Murray T (2006) Multivariate controls on the incidence of glacier surging in the Karakoram Himalaya. Arct Antarc Alp Res 38(4):489–498

Kamb B (1985) Glacier surge mechanism: 1982–1983-surge of variegated glacier, Alaska. Am Assoc Adv Sci Pub 227(4686):469–479

Flint RF (1971) Glacial and quaternary geology. Wiley, New York, 892

Hambrey MJ, Dowdeswell JA (1997) Structural evolution of a surge-type glacier: Hessbreen, Svalbard. Ann Glaciol 24:375–381

Harrison WD, Post AS (2003) How much do we really know about glacier surging? Ann Glaciol 36:1–6

Hewitt K (1969) Glacier surge in the Karakorum Himalaya (Central Asia). Can J Earth Sci 6:1009–1018

Hewitt K (2007) Tributary glacier surges: an exceptional concentration at Panmah glacier, Karakoram Himalaya. J Glaciol 53(181):181–188

Raina VK, Srivastava D (2008) Surging glacier in the Himalaya. Glacier Atlas of India, Geological Society of India, Special Publication, vol 287, pp. 51–58

Chandra R, Tangri AK (2011) Monitoring dynamics of Chong Kumdan surging glaciers, Shyok valley, Karakoram Himalaya, Ladakh region, Jammu & Kashmir, India using remote sensing and GIS techniques. Proceedings of the national seminar on late quaternary geology of the Himalayan orogen and foreland basin, Lucknow, pp 57–59

SAC-ISRO Technical Report (2010) Snow and glaciers of the Himalayas-Inventory and monitoring, vol 88, pp. 268

Tangri AK, Chandra R, Yadav SKS (2011) Assessing the impact of climate change on surging glaciers. A case study in Shyok valley of Karakoram Himalayas, J & K State, India. Proceedings of the national conference on science of climate change and earth's sustainability, Lucknow

Tangri AK, Chandra R, Yadav SKS (2008). Evidences of surging glaciers in Shyok valley, Karakorum Himalaya, Jammu & Kashmir India – the first cut results. Proceedings of the national seminar on glacial geomorphology and paleoglaciation in Himalaya March, Lucknow, pp 58–59

Wake CP, Searle MP (1993) Rapid advance of the Pumarikish glacier, Karakoram Himalaya. J Glaciol 39:204–206

Chapter 5
Antarctic Climate Variability During the Past Few Centuries Based on Ice Core Records from Coastal Dronning Maud Land and Its Implications on the Recent Warming

Meloth Thamban, Sushant S. Naik, C.M. Laluraj, Arun Chaturvedi, and Rasik Ravindra

5.1 Introduction

The modern climate over the Antarctica and the Southern Ocean is an outcome of the interplay of the ice sheet, ocean, sea ice, and atmosphere and their response to past and present climate forcing. With ~98% of its area covered with snow and ice, Antarctic continent reflects most of the sun's light rather than absorbing it. This in association with a combination with various other factors like extreme dryness of its atmosphere and higher elevation make Antarctica the coldest region on Earth. The large temperature difference between the equator and pole leads to poleward transport of heat which, in association with the Coriolis force imposed by the Earth's rotation, leads to the development of Hadley Cell, a large scale global circulation system (Trenberth and Caron 2001). The related pressure gradient at the Earth's surface between subtropics (high pressure) and polar belt (low pressure) forces air to move eastwards under the influence of the Coriolis Force, creating the mid-latitude westerlies that drive the world's largest and strongest current system, the Antarctic Circumpolar Current (Pickard and Emery 1990; Turner et al. 2009). As a result of the intense radiative cooling during the winter, there exists a strong horizontal temperature gradient between the continent and the ocean with a strong vertical temperature

M. Thamban (✉) • C.M. Laluraj • R. Ravindra
National Centre for Antarctic and Ocean Research, Headland Sada, Vasco-da-Gama, Goa 403004, India
e-mail: meloth@ncaor.org

S.S. Naik
National Institute of Oceanography, Dona Paula, Goa 403004, India

A. Chaturvedi
Glaciology Division, Geological Survey of India, Aliganj, Lucknow, Uttar Pradesh 226024, India

R. Sinha and R. Ravindra (eds.), *Earth System Processes and Disaster Management*, Society of Earth Scientists Series 1, DOI 10.1007/978-3-642-28845-6_5,

gradient (Van den Broeke 2000). The north-to-south distribution of surface pressure around Antarctica is subject to remarkable variability in the intensity of the meridional pressure gradient and its zonal location. Due to the circumpolar nature of this variation, it is called the Southern Annular Mode (SAM), which is the principal mode of variability in the atmospheric circulation of the southern extratropics and high latitudes (see Trenberth et al. 2007). The SAM has a zonally symmetric or annular structure, with synchronous anomalies of opposite sign in high latitudes and midlatitudes (Lefebvre et al. 2004). The El Niño Southern Oscillation (ENSO) on the other hand, is the farthest reaching climatic cycle on Earth on decadal and sub-decadal time scales, affecting even the Southern Hemisphere (see Bromwich et al. 2000). The ENSO signals can be identified in the physical and biological environment of the Antarctic, although some of the links are not very robust and there can be large differences in the extra-tropical response to near-identical events in the tropics (see Turner 2004 for a review).

The Antarctic climate system fluctuates widely on sub-annual to millennial time scales, in tandem with the global climate system. Since the time-series observations of climatic parameters were initiated only from the International Geophysical Year (1957–1958), most of the instrumental records of Antarctic climate are only some decades old. Based on the available time series data, Antarctica seems to have undergone complex and significant temperature changes in recent decades (Turner et al. 2005, 2006). While the largest annual warming trends are found on the western and northern parts of the Antarctic Peninsula (e.g., +0.56°C per decade at Faraday/Vernadsky station), many stations in East Antarctica and Amundsen-Scott Station at the South Pole in particular has shown a small cooling in the annual mean temperature of −0.05°C per decade (Turner et al. 2009).

Notwithstanding these studies to understand the climate change in Antarctica, the spatial and temporal complexity of Antarctic climate are still poorly understood because of the limited and short periods of observational and instrumental data. Analysis of ice core records offer one of the most direct and accurate method to study the Antarctic climate change beyond the instrumental limits (Schneider et al. 2006; Mayewski et al. 2009). Ice core records from Polar Regions present continuous and highly resolved long-term records of reliable information on major atmospheric parameters like temperature, composition and trace gases. Among the various proxy variables used, the stable isotope ratios of oxygen ($\delta^{18}O$) and hydrogen (δD) offer the most critical information on the past changes in temperature. Owing to the differences in saturation vapour pressure and molecular diffusivity, isotopic fractionation takes place at each phase change of atmospheric water (Dansgaard 1964). Since the heavy isotopes in precipitation decrease with the temperature of condensation, stable isotope ratios in ice cores provide quantitative proxy records of past temperature. However, factors such as temperature of the source from which moisture evaporated, subsequent cycles of condensation and evaporation, change in moisture source region and seasonality of precipitation make the interpretation of temperature from water isotopes difficult (Jouzel et al. 2003). Some of these factors could be resolved using a second order isotopic parameter called deuterium excess ($d = \delta D - 8 * \delta^{18}O$) which is calculated from the combined measurements of δD and $\delta^{18}O$. This 'deuterium excess'

(Dansgaard 1964) depends on the temperature and relative humidity of the evaporative source, which in turn contains information about conditions prevailing in these source regions (Jouzel et al. 2003).

Isotope records of ice cores have indeed provided insight into the tropical-polar teleconnections (Isaksson and Karlén 1994; Schneider et al. 2006; Naik et al. 2010a, b). A continental wide network of ice core records with annually resolved proxy data that focus on the past few centuries are most fundamental for the study of recent climate change (Mayewski et al. 2005a). The international effort in this regard was pioneered by the International Trans Antarctic Scientific Expedition (ITASE) that established calibration tools as well as reconstruction of climate indices based on shallow ice cores covering the past ~200 years (Mayewski et al. 2005a, 2009; Bertler et al. 2006). The increased focus on the annually-dated shallow ice cores has indeed improved our understanding on the recent climate variability in Antarctica and forcing mechanisms. The inherent complexities in the proxy records also necessitate the availability of a large array of well-dated firn/ice cores from the coastal regions of Antarctica. With this backdrop, the National Centre for Antarctic and Ocean Research (NCAOR) has taken up studies on snow and shallow ice core from the central Dronning Maud Land (DML) in East Antarctica. The initial results suggest that understanding the modern biogeochemical processes, careful selection of core sites based on glaciological and geophysical survey, as well as critical analysis and interpretation of proxy parameters would be crucial for reliable climatic interpretations based on ice cores (Nijampurkar et al. 2002; Thamban et al. 2006, 2010; Laluraj et al. 2009, 2010; Marshall et al. 2009; Antony et al. 2010; Naik et al. 2010a, b). The present study examines the stable isotope records of two ice cores (IND-22/B4 and IND-25/B5) from the coastal Dronning Maud Land (DML) and provides improved understanding on the climatic and environmental variability in this part of Antarctica during the past few centuries.

5.2 Material and Methods

The present study of climatic reconstruction in coastal Antarctica utilizes two ice core records from the coastal Dronning Maud Land region in East Antarctica (Fig. 5.1). Among these, one core (IND-22/B4, length: 62 m) was collected near the Tallaksenvarden nunatak (location: 70°51.3′S and 11°32.2′E) during the 22nd Indian Scientific Expedition to Antarctica (InSEA) and represent a low altitude (elevation: 680 m), low accumulation (accumulation: up to ~170 kg m^{-2} a^{-1}) region. The second core (IND-25/B5; length: 65 m) was collected during the 25th InSEA near the Humboldt Mountains (location: 71° 20′S and 11° 35′E) and come from higher altitude (elevation: 1,300 m) and higher accumulation (accumulation rate: up to 528 kg m^{-2} a^{-1}) site (Fig. 5.1).

The ice core drilling was carried out using an electromechanical drill system (GeoTecs Ltd, Japan; Model: Type D-2) during the austral summers (Fig. 5.2). The retrieved ice cores (diameter 10 cm) were wrapped in labelled low density

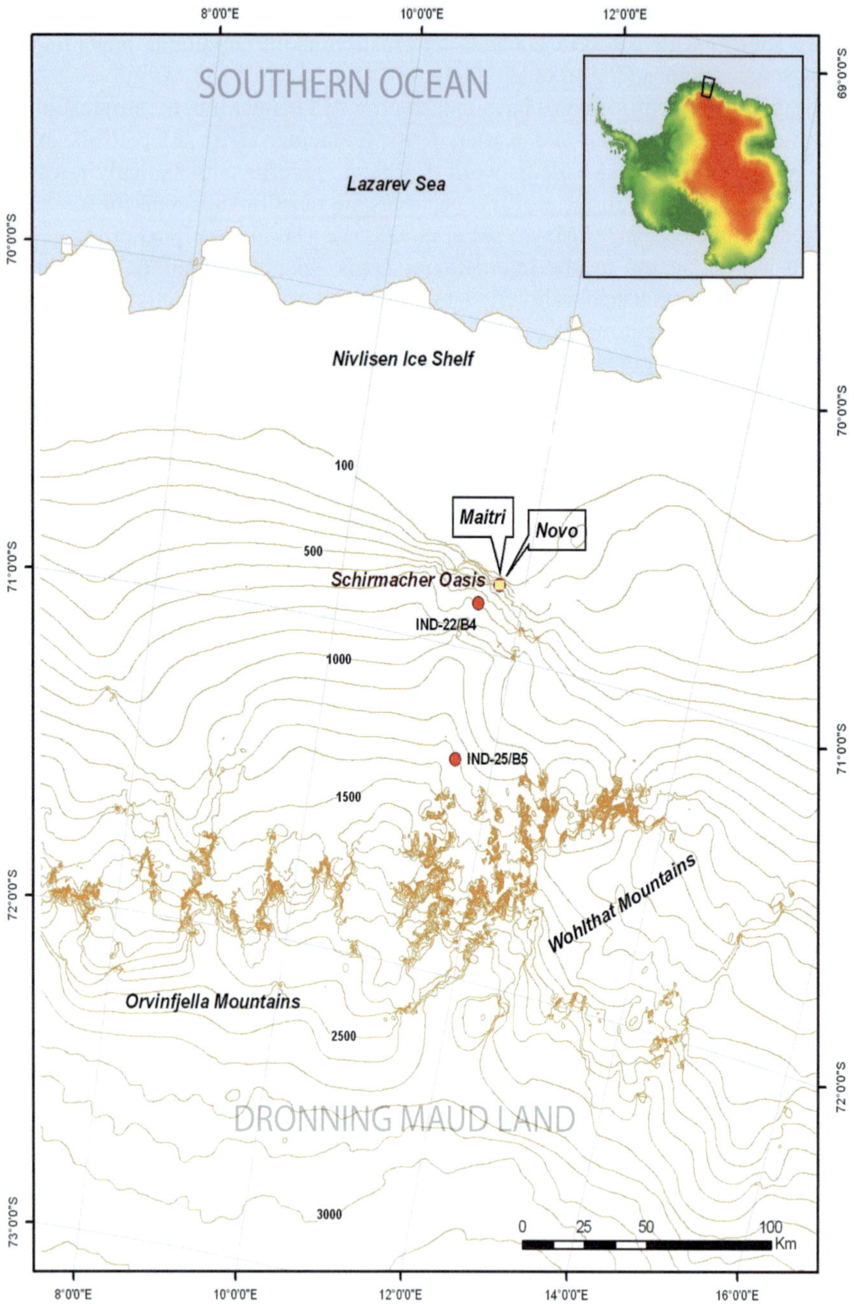

Fig. 5.1 Study area and locations of ice cores IND-22/B4 and IND-25/B5 within the Dronning Maud Land

Fig. 5.2 Field photographs of the Indian ice core project and the laboratory activities at NCAOR. (**a**) Ground Penetrating Radar survey for site location; (**b**) ice core drilling in progress; (**c**) the drill set up; (**d**) team with the first ice core, (**e**) ice cores ready for transport; (**f**) samples processing at −15°C cold room facility; (**g**) clean room facility and ion chromatography system; (**h**) isotope ratio mass spectrometer facility

polyethylene (LDPE) rolls and subsequently shipped using custom-made expanded polypropylene (EPP) insulated containers stored in −20°C deep freezer facilities. These cores were archived at the NCAOR cold room facility (−20°C), until further processing in the laboratory. Processing of the ice core samples were carried out at the ice core processing facility at NCAOR. The ice core samples were first sub-sampled using custom-made bandsaw machines at the cold room (−15°C) processing facility. Subsequently, the selected samples were manually decontaminated by removing the outer layer using microtome blades. Prior to analyses, samples were thawed at room temperature in a Class 100 clean room.

Stable isotope ratios of the ice core samples were measured following standard methods using an 'Isoprime' (GV instruments) Isotope Ratio Mass Spectrometer. The oxygen isotope ratios ($^{18}O/^{16}O$) were determined using the method of equilibrating the water samples with CO_2 gas following the procedure of Epstein and Mayeda (1953); for hydrogen isotope ratio (D/H) analysis, the method of H_2–H_2O equilibration using platinum catalysts was followed (Coplen et al. 1991). All measurements were made against IAEA (International Atomic Energy Agency) reference material Vienna Standard Mean Ocean Water (V-SMOW) and Standard Light Antarctic

Precipitation (SLAP) from National Institute of Standards and Technology (NIST). The isotopic composition is expressed in per mil (‰) deviation of the sample from V-SMOW standard. The external precision (1σ) obtained for $\delta^{18}O$ and δD analyses using daily measurements of a calibrated laboratory standard (CDML1; snow from a coastal DML site) were ±0.05‰ and ±0.77‰ respectively, based on 20 repeat measurements. For IND-22/B4 ice core, a total of 225 sub-samples were studied within the top ~50 m length representing the last ~470 years. Compared to this, 1,300 samples were studied in the 65 m long IND-25/B5 core, leading to an ultra-high resolution time series data representing the last ~100 years.

Chronological model for the IND-22/B4 core was achieved by volcanic chronology using the anomalies in non-sea-salt sulphate ($nssSO_4^{2-}$) records (Thamban et al. 2006). The detailed chronological reconstruction of the core were made using the sulphate anomalies globally ascribed to some of the major volcanic eruptions, including the most prominent and unique doublet peaks ascribed to Tambora (1815) and Unknown (1809) eruptions. Accordingly, the ~50 m long section of the ice core represented the past 470 years with an estimated chronological error of ±3 years. The age control for the IND-25/B5 ice core was based on multiple and complimentary methods: (1) annual layer determination using the summer maxima in $\delta^{18}O$ values; (2) $nssSO_4^{2-}$ (non-sea salt sulphate) markers of volcanic eruptions; and (3) Tritium markers (Naik et al. 2010b). The exceptionally high annual accumulation rates at this site provided high sampling resolution (11–14 samples per year) throughout the core. A five point smoothing was used for the entire $\delta^{18}O$ profile so as to facilitate the summer layer counting. Chronology thus determined by counting the seasonal cycles in the oxygen isotope data was further supported by the identification of well-known volcanic eruptions of Pinatubo (AD 1991) and Azul (AD 1932). Further to validate the chronology, the tritium spikes originated from atomic bomb testing activities during the early 1960s were also measured from a selected section in the core. Such rigorous and complimenting dating procedures provided an excellent chronological framework for the IND-25/B5, with the 65 m core representing that last 100 years (1905–2005) with an error of ±2 years.

5.3 Results

Core IND-25/B5: The $\delta^{18}O$ record varied between −20‰ and −37‰ whereas the δD values varied between −150‰ and −293‰ (Fig. 5.3). The isotopic composition of worldwide freshwater sources are defined by the Global Meteoric Water Line (GMWL) with the relation $\delta D = 8 * \delta^{18}O + 10$ (Craig 1961). For the entire 65 m length of the IND-25/B5 ice core presented here, the relationship between the oxygen and hydrogen isotopes followed the GMWL as defined by $\delta D = 8.2 * \delta^{18}O + 10.4$ ($r^2 = 0.9$). The isotope records revealed relatively more negative isotope values during 1930–1950 AD as well as large amplitude fluctuations for the past few decades (Fig. 5.3). The Southern Annular Mode (SAM) index data used

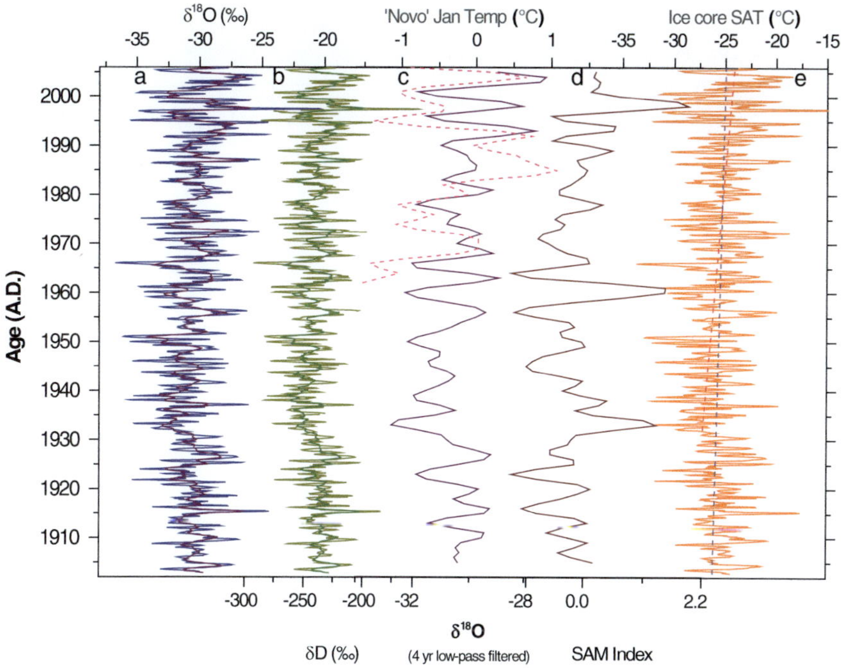

Fig. 5.3 Down core profiles of IND-25/B5 core: (**a**) $\delta^{18}O$ record with annual smoothing; (**b**) δD record with annual smoothing; (**c**) interrelationships between the 4 year low pass filtered $\delta^{18}O$ data with the January temperatures from Novolazarevskaya station (till 1964); (**d**) Southern Annular Mode (*SAM*) index profile; (**e**) estimated surface air temperature (*SAT*) record, showing a warming trend of ~1°C during 1905–2005 (*thin dashed line*) and enhanced warming by ~3°C since 1930s (*thick dashed line*)

in this study is a reconstructed index based on multiple linear regressions of the meteorological data from research stations within the Southern Hemisphere (Jones et al. 2009; Naik et al. 2010b). The Southern Oscillation Index (SOI) calculated from the monthly or seasonal fluctuations in the air pressure difference between Tahiti and Darwin (data from Australian Bureau of Meteorology) is used for monitoring ENSO.

Spatial variations of oxygen isotopes in polar snow/ice have been used to determine the empirical relationship between the mean annual temperature and the $\delta^{18}O$ content in snow (e.g., Isaksson and Karlén 1994; van Ommen and Morgan 1997; Masson-Delmotte et al. 2008). The close similarity between the 'Novo' summer temperature record (from 1962 onwards) and the corresponding periods in IND-25/B5 $\delta^{18}O$ record suggest that the $\delta^{18}O$ record could be utilised for deducing the general surface air temperature (SAT) for this region. The temporal $\delta^{18}O$-T calibration studies in the western Dronning Maud Land (DML) provided spatial slopes ranging from 1.16‰ to 1.31‰ change in $\delta^{18}O$/°C with similar high values for eastern DML (Isaksson and Karlén 1994). Since the relationship between $\delta^{18}O$ and temperature are not well understood in the core sites, the $\delta^{18}O$-T slope

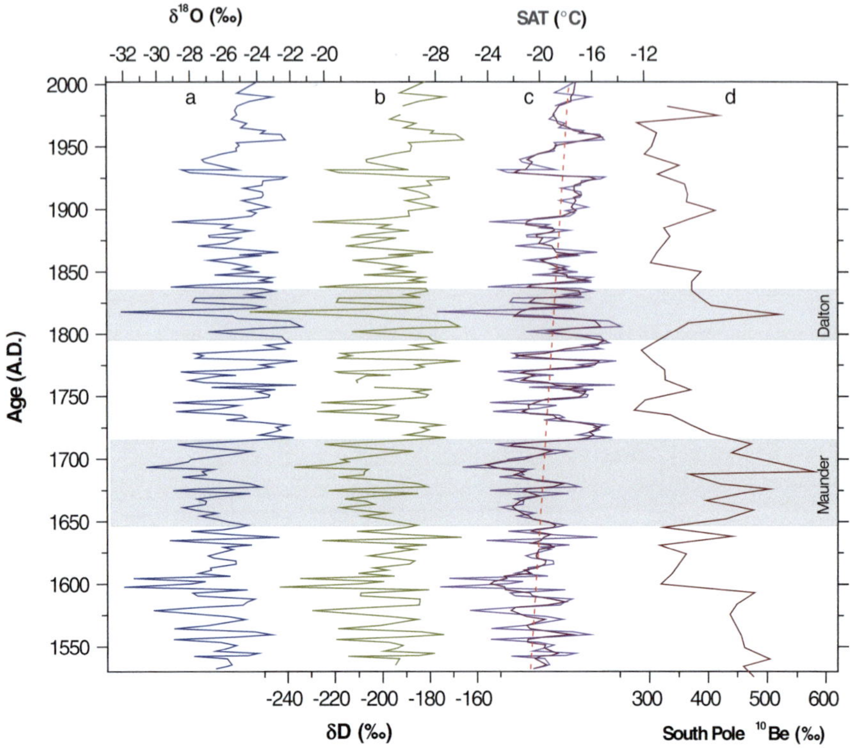

Fig. 5.4 Down core records of IND-22/B4 core: (**a**) $\delta^{18}O$ profile; (**b**) δD profile; (**c**) surface air temperature (*SAT*) profile along with the long term warming trend (*dashed line*); (**d**) South Pole ^{10}Be proxy data of solar variability (Bard et al. 2000), showing the relationship between the Antarctic temperature and solar activity. The *shaded* intervals are periods of reduced solar activity events

was derived using data from central DML region based on the regional study of Masson-Delmotte et al. (2008). The $\delta^{18}O$-T calibration provides a spatial slope of 1.31‰/°C for this region, which is in agreement with other studies from DML region (see Naik et al. 2010b). Based on this spatial slope, the average SAT thus derived for IND-25/B5 was −25.6°C (1905–2005), with the temperatures fluctuated between −33.7°C and −12.4°C (Fig. 5.3).

Core IND-22/B4: Compared to the IND-25/B5 core, the $\delta^{18}O$ and δD profiles of IND-22/B4 core provided longer time series, but with reduced temporal resolution (Fig. 5.4). The $\delta^{18}O$ and δD records of this core revealed significant variability on a annual to decadal scale with a major shift in $\delta^{18}O$ (by ~8‰) to more positive values around 1710 AD (Fig. 5.4). The $\delta^{18}O$ values varied between −32.2 and −21.2‰ (mean −25.5‰) and the δD values varied between −256.2‰ and −157.9‰ (mean −196.4‰). The $\delta^{18}O$ and δD records of IND-22/B4 revealed similar temporal fluctuations with a deviation from the GMWL given by $\delta D = 7.5 * \delta^{18}O - 5.5$ ($r^2 = 0.86$). In addition to the high amplitude multi-annual fluctuations, the $\delta^{18}O$

and δD records revealed a long-term systematic increase towards positive values from older to modern times (Fig. 5.4). The estimated surface air temperature (SAT) using the δ^{18}O-T relationship as defined earlier showed that during the last ~470 years, the temperature fluctuated between −27.9°C and −13.8°C at the core site (Fig. 5.4). The record showed a long term increase towards recent times with two major cold spells during ~1640–1710 AD and ~1790–1830 AD.

5.4 Discussion

Considering that the two ice cores studied here are spatially and temporally different, these could provide valuable insights on the recent climate change in the broad coastal regions of East Antarctica. In order to use the δ^{18}O record as proxy for surface air temperatures at the site, we also compared the δ^{18}O record of the IND-25/B5 core with the measured air temperature from this region (Fig. 5.3). Relatively longer (~50 years) temperature data for this region is only available from the Russian station 'Novolazarevskaya' (Novo) (70° 46′S and 11° 50′ E) which, is located at a much lower altitude (elevation ~100 m a.s.l) and ~65 km north of the core site. Although the 'Novo' temperature record may only be a broad representation of the conditions prevailing at the core site, a weak positive correlation ($r = 0.15$) was found between annual ice core δ^{18}O and annual 'Novo' surface air temperatures, with substantially enhanced correlations on decadal scales. This suggests that while processes other than temperature do influence the annual isotopic profile, the long-term oscillations in δ^{18}O record could be used as a climatic proxy indicator. Accordingly, we discuss below the stable isotope records of the two ice cores in terms of coastal Antarctic variability vis-a-vis the global climatic modes.

5.4.1 Stable Isotope Records of Antarctic Climate Variability and Global Teleconnections

The considerable variations in the δ^{18}O and δD records as well as the accumulation rates on an interannual to decadal scale seems to be associated with circulation changes in this region which could be linked to low and mid latitude climatic modes. The El Niño–Southern Oscillation (ENSO) is one such phenomenon which leads a pattern of warm and cold sea surface temperature (SST) anomalies in the central and eastern equatorial Pacific with coupled atmospheric changes. Although the ENSO is dominant in the Pacific, its signals are well recorded in remote regions of the world (Yuan 2004; Fogt and Bromwich 2006; Gregory and Noone 2008; Naik et al. 2010b). This teleconnection occurs through the Pacific South American pattern (PSA), which represents a series of positive and negative geopotential height anomalies initiated from tropical convection and extending from central equatorial Pacific to Australia, South Pacific near Antarctica, South America and then bending northwards towards Africa (Turner 2004). During the ENSO events, the PSA pattern gives rise to geopotential height anomalies over the Amundsen-Bellingshausen seas and the Weddell Sea region. Such an Antarctic Dipole phenomenon is commonly observed in the out-of-phase relationship in sea ice and surface temperature anomalies in South Pacific and South Atlantic (Yuan 2004).

Studies have revealed that the Southern Annular Mode (SAM) significantly influences the Antarctic climate (Kwok and Comiso 2002). The SAM, a hemisphere-wide atmospheric pattern, contributes to ~35% of variance in sea level pressure or geopotential height on a large range of time scales (Marshall 2003). Several studies emphasised the crucial role of SAM on the recent climate changes across the Antarctic continent. Over the last 50 years, the SAM has shifted more into its positive phase with decreases of surface pressure over the Antarctic and corresponding increases at mid-latitudes (Marshall 2003). The positive trend in the SAM seems to have resulted in a strengthening of the circumpolar westerlies and contributed to the spatial variability in Antarctic temperature change (Kwok and Comiso 2002; Schneider et al. 2004; Marshall 2007). Most importantly, ENSO is also known to affect the SAM in a highly non-linear way. The two forcings can combine, partially offset or even enhance their influence on each other and the Southern Hemisphere as a whole (Bertler et al. 2006; Fogt and Bromwich 2006; Gregory and Noone 2008; Divine et al. 2009; Naik et al. 2010a, b).

The existence of SAM in instrumental and proxy records and its extent was evaluated through a detailed study of the regional atmospheric circulation changes associated with a reversal in the sign of the relationship between the SAM and near-surface temperatures in coastal parts of East Antarctica. This study based on instrumental data from Halley station as well as the IND-25/B5 proxy record from coastal Dronning Maud Land revealed that the key factor affecting the regional SAM-temperature relationship is the relative magnitude of two climatological low pressure centres to the west and east of the area, which determines the source region of air masses advected into the locality (Marshall et al. 2009). A similar reversal is observed in the sign of the correlation between the SAM and oxygen-isotope values from the IND-25/B5 ice core from Dronning Maud land, located some 1,200 km east of Halley. This relationship examined throughout the twentieth century, by comparing the isotope data with SAM reconstruction profiles, demonstrates marked decadal variability. It is suggested that the switches in the SAM-temperature relationship are more likely to reflect natural variability in the long-wave patterns over the Southern Ocean rather than the influence of an anthropogenic forcing (Marshall et al. 2009).

Comparison of the ice core $\delta^{18}O$ record with the reconstructed index of SAM for the period 1905–2005 was made after a 4-year low pass filtering of both data (Fig. 5.3). Such a filtering could take care of the well known 4–5 years variability within the SAM data (Thompson and Wallace 2000). Although the correlations are low ($r^2 = -0.01$; $p = 0.1$), it is negative with high SAM corresponding to lower $\delta^{18}O$ and vice versa. This suggests that higher polarity of SAM corresponds to lower temperatures and vice versa. This is a component of the opposite relationship driven by SAM on surface air temperature over the west Antarctica (Gregory and Noone 2008) and east Antarctica. SAM is also known to play an important role in driving decadal temperature changes in the Southern Hemisphere. A running decadal correlation between the annual $\delta^{18}O$ and SAM records also indicated an overall negative relationship, with certain periods (1918–1927; 1938–1947; and 1989–2005) showing absence of correlations or even weak positive relationships (Naik et al. 2010b).

Relationships between the surface air temperature, SAM and ENSO in the study region were established using the ice core annual $\delta^{18}O$ data, the reconstructed SOI as well as the SAM indices. The analysis showed that for the period wherein the $\delta^{18}O$ and SAM relationship was insignificant or positive (1918–1927; 1938–1947 and 1989–2005), the relationships between the SOI and SAM are also in-phase (Naik et al. 2010b). This indicates that during warm ENSO events, positive $\delta^{18}O$ anomalies were noticed in the study site and vice versa. Such a relation between SOI and $\delta^{18}O$ was also reported from the coastal DML (Divine et al. 2009). Interestingly, the above periods (1918–1927; 1938–1947 and 1989–2005) incorporate several El Niño and La Niña years. When the years of El Niño and La Niña events are omitted from the records, the relationship between the filtered $\delta^{18}O$ and SAM data became statistically significant ($r^2 = -0.2$; $p = 0.02$), suggesting that ENSO events weakens the SAM-temperature relationship during these periods (Naik et al. 2010b). The results thus reveals that during the past century, the combined influence of ENSO-SAM modes have controlled the temporal changes in $\delta^{18}O$ values at the core site.

The estimated air temperatures (SAT) of IND 22/B4 revealed that the cooler temperatures prevailed during 1550–1715 AD (Fig. 5.4). This interval is coeval with the so called 'Little Ice Age' (LIA) widely depicted in the northern hemisphere and elsewhere. Many researchers consider the LIA a global event and demonstrated the existence of this cool interval in Antarctic proxy records (e.g., Kreutz et al. 1997). However, the timing and signatures of LIA in Antarctica is controversial (Bradley and Jones 1993; Bard et al. 2000). An important question is how this fits in the proposed anti-phased warming ('bipolar seesaw') in southern hemisphere during the northern hemisphere LIA cooler conditions (Broecker 2000). Another interesting aspect of $\delta^{18}O$ variability is the more negative values (enhanced cooling) during periods of reduced solar activity like the Dalton Minimum (~1790–1830 AD) and Maunder Minimum (~1640–1710 AD), indicating relatively lower air temperatures during these periods. These were the periods when the sunspots were rare or absent, leading to the decrease in solar irradiance due to a decrease in magnetic disturbances in the convective plasma near the Sun's surface (Bard et al. 2000). Climatic reconstructions using a variety of records have indeed suggested that solar forcing on the global climate system in addition to other natural and anthropogenic factors (Bard et al. 2000; Mayewski et al. 2005b). A possible mechanism for the solar modulation of the southern hemisphere climate system is through the dynamical coupling between the troposphere and stratosphere during the late spring/winter (Kuroda et al. 2007). Accordingly, enhanced solar activity would enhance the ultraviolet and ozone production that could lead to stronger troposphere-stratosphere coupling through a strong interaction between the planetary waves and radiation. The present study in association with proxy records of solar variability in Antarctica thus suggest a possible influence of solar forcing on the Antarctic climate change during the past few centuries.

5.4.2 Reconstructed Surface Air Temperature Variability and Its Implications on Recent Warming

Application of the δ^{18}O-T spatial slope (1.31‰/°C) to the δ^{18}O profiles of IND-25/B5 and IND-24/B4 facilitated the estimation of surface air temperature in the study area (Fig. 5.4). The SAT record of IND-25/B5 exhibited high amplitude oscillations, with an average warming of 1°C for the entire century (1905–2005) with a trend of 0.1°C/10 years. However, the data revealed greatly enhanced warming trend of ~3°C during 1930–2005, with a warming of ~0.4°C/10 years. Based on the available instrumental data, it was demonstrated that while the western and northern parts of the Antarctic Peninsula is significantly warming, the East Antarctica in general and the South Pole in particular has shown a small cooling (Turner et al. 2005, 2006). However, a continental wide reconstruction for the last 50 years revealed that the significant warming cover most of the West Antarctica (0.1°C/10 years) and even part of the East Antarctica (Steig et al. 2009). Analysis of instrumental records from the coastal East Antarctic 'Novo' station also indicate a warming for the last ~50 years with similar temperature trend as found in the IND-25/B5 ice core (Naik et al. 2010b). Further support for such a warming derive from another coastal ice core from DML, which estimates a positive trend of the order of 0.12‰/decade for a similar time period (Divine et al. 2009). Therefore, both the instrumental data as well as the proxy records confirms a significant warming during the past century at coastal DML region.

Compared to the above, the SAT record of IND-22/B4 revealed an estimated warming of 3°C for the past 470 years (Fig. 5.4). Although this warming trend is lower than at the IND-25/B5 site, it may be noted that the relatively longer time period as well as the coarser sampling resolution compared to the IND-25/B5 core may introduce more errors in the estimation. However, it is significant that both the records confirm that the region has indeed warmed during the past few centuries. These records while confirming the instrumental record of recent warming at Novo refute the observed absence of any trend at the Syowa station (coastal East Antarctica) and the observed slight cooling at Amundsen-Scott Station (South Pole). Considering such heterogeneity in the past climatic conditions, the current estimation of temperature trends across the East Antarctica based on extrapolations using few station records (e.g., Chapman and Walsh 2007), needs to be vigorously tested. It is suggested that spatially distributed ice core derived SAT profiles could provide valuable data in filling the large gaps as well as extending the climatic records in Antarctica.

5.5 Conclusions

Analysis of the available instrumental records revealed that Antarctica has undergone complex and significant temperature changes during the recent decades, with a substantial warming in most parts of the West Antarctica and no such trends in East Antarctica. In order to evaluate the climate variability beyond the instrumental records at the coastal Dronning Maud Land (East Antarctica), stable isotope proxy records of two ice core records were examined. The variations in the $\delta^{18}O$ and δD records in two cores correspond to changes in low and mid latitude climatic modes like the Southern Annular Mode (SAM) and El Niño Southern Oscillation (ENSO). The estimated surface air temperatures using the $\delta^{18}O$ profiles of two ice cores revealed significant warming by 0.6–1°C per century during the past five centuries. This study corroborates that the coastal Dronning Maud Land (East Antarctica) have been experiencing significant warming during past centuries and trend is alarmingly increasing (~0.4°C per decade) in the recent decades.

Acknowledgements We thank the Ministry of Earth Sciences (New Delhi) for the financial support. We also acknowledge the excellent field efforts of Dr. A. Rajakumar, A. Singh, P. C. George as well as the laboratory efforts of B. L. Redkar. The scientific and logistic team members of the 22nd and 25th Indian Scientific Expedition to Antarctica are thanked for their cooperation and support. This is NCAOR Contribution No. 20/2012.

References

Antony R, Thamban M, Krishnan KP, Mahalinganathan K (2010) Is cloud seeding in coastal Antarctica linked to biogenic bromine and nitrate variability in snow? Environ Res Lett 5. doi:10.1088/1748-9326/5/1/014009

Bard E, Raisbeck G, Yiou F, Jouzel J (2000) Solar irradiance during the last 1200 years based on cosmogenic nuclides. Tellus 52:985–992

Bertler NAN, Naish TR, Kiphstuhl S, Barrett PJ, Mayewski PA, Kreutz KJ (2006) The effects of joint ENSO–Antarctic Oscillation forcing on the McMurdo Dry Valleys, Antarctica. Antarct Sci 18(4):507–514

Bradley RS, Jones PD (1993) "Little Ice Age" summer temperature variations: their nature and relevance to recent global warming trends. Holocene 3:367–376

Broecker WS (2000) Was a change in thermohaline circulation responsible for the Little Ice Age? Proc Natl Acad Sci 97:1339–1342

Bromwich DH, Rogers AN, Kallberg P, Cullather RI, White JWC, Kreutz KJ (2000) ECMWF analysis and reanalysis depiction of ENSO signal in Antarctic precipitation. J Climate 13:1406–1420. doi:10.1175/1520-0442(2000) 013

Chapman WL, Walsh JE (2007) A synthesis of Antarctic temperatures. J Climate 20:4096–4117

Coplen TB, Wildman JD, Chen J (1991) Improvements in the gaseous hydrogen-water equilibration technique for hydrogen isotope ratio analysis. Anal Chem 63(9):910–912

Dansgaard W (1964) Stable isotopes in precipitation. Tellus 16:436–468

Divine DV, Isaksson E, Kaczmarska M, Godtliebsen F, Oerter H, Schlosser E, Johnsen SJ, van den Broeke M, van de Wal RSW (2009) Tropical Pacific–high latitude south Atlantic teleconnections as seen in $\delta^{18}O$ variability in Antarctic coastal ice cores. J Geophys Res 114: D11112. doi:10.1029/2008JD010475

Epstein S, Mayeda T (1953) Variations of O^{18} contents of waters from natural sources. Geochim Cosmochim Acta 4:213

Fogt RL, Bromwich DH (2006) Decadal variability of the ENSO teleconnection to the high-latitude south Pacific governed by coupling with the southern annular mode. J Climate 19:979–997

Gregory S, Noone D (2008) Variability in the teleconnection between the El Niño-Southern Oscillation and West Antarctic climate deduced from West Antarctic ice core isotope records. J Geophys Res 113:D17110. doi:10.1029/2007JD009107

Isaksson E, Karlén W (1994) Spatial and temporal patterns in snow accumulation, western Droning Maud Land, Antarctica. J Glaciol 40(135):399–409

Jones JM, Fogt RL, Widmann M, Marshall GJ, Jones PD, Visbeck M (2009) Historical SAM variability. Part I: century length seasonal reconstructions. J Climate. doi:10.1175/2009JCLI2785.1

Jouzel J, Vimeux F, Caillon N, Delaygue G, Hoffmann G, Masson-Delmotte V, Parrenin F (2003) Magnitude of isotope/temperature scaling for interpretation of central Antarctic ice cores. J Geophys Res 108(D12):4361. doi:10.1029/2002JD002677

Kreutz KJ, Mayewski PA, Meeker LD, Twickler MS, Whitlow SI, Pittalwala II (1997) Bipolar changes in atmospheric circulation during the Little Ice Age. Science 277:1294–1296

Kuroda Y, Deushi M, Shibata K (2007) Role of solar activity in the troposphere-stratosphere coupling in the Southern Hemisphere winter. Geophys Res Lett 34:L21704. doi:10.1029/2007GL030983

Kwok R, Comiso JC (2002) Southern ocean climate and sea ice anomalies associated with the Southern Oscillation. J Climate 15:487–501

Laluraj CM, Krishnan KP, Thamban M, Rahul M, Chaturvedi A, Naik SS, D'Souza W, Ravindra R (2009) Origin and characterisation of microparticles in an ice core from the Central Dronning Maud Land, East Antarctica. Environ Monit Assess 149(1–4):377–383

Laluraj CM, Thamban M, Naik SS, Redkar BL, Chaturvedi A, Ravindra R (2010) Nitrate records of a shallow ice core from East Antarctica: atmospheric processes, preservation and climatic implications. Holocene 21:351–356. doi:10.1177/0959683610374886

Lefebvre W, Goosse H, Timmermann R, Fichefet T (2004) Influence of the Southern Annular Mode on the sea ice-ocean system. J Geophys Res 109:C09005. doi:10.1029/2004JC002403

Marshall GJ (2003) Trends in the southern annular mode from observations and reanalyses. J Climate 16:4134–4143

Marshall GJ (2007) Half-century seasonal relationships between the Southern Annular Mode and Antarctic temperatures. Int J Climatol 27:373–383

Marshall GJ, Battista S, Naik SS, Thamban M (2009) Analysis of a regional change in the sign of the SAM-temperature relationship in Antarctica. Clim Dyn 36:277–287. doi:10.1007/s00382-009-0682-9

Masson-Delmotte V et al (2008) A review of Antarctic surface snow isotopic composition: observations, atmospheric circulation, and isotopic modelling. J Climate 21:3359–3387

Mayewski PA, Maasch KA, White JWC, Meyerson E, Goodwin I, Morgan VI, van Ommen T, Curran MAJ, Souney J, Kreutz K (2004) A 700 year record of Southern Hemisphere extra-tropical climate variability. Ann Glaciol 39:127–132

Mayewski PA, Frezzotti M, Bertler N, Van Ommen T, Hamilton G, Jacka TH, Welch B, Frey M, Dahe Q, Jiawen R, Simöes J, Fily M, Oerter H, Nishio F, Isaksson E, Mulvaney R, Holmund P,

Lipenkov V, Goodwin I (2005a) The International Trans-Antarctic Scientific Expedition (ITASE): an overview. Ann Glaciol 41(1):180–185

Mayewski PA, Maasch KA, Yan Y, Kang S, Meyerson EA, Sneed SB, Kaspari SD, Dixon DA, Osterberg EC, Morgan VI, Ommen TV, Curran MAJ (2005b) Solar forcing of the polar atmosphere. Ann Glaciol 41:147–154

Mayewski PA, Meredith MP, Summerhayes CP, Turner J, Worby A, Barrett PJ, Casassa G, Bertler NAN, Bracegirdle T, Naveira Garabato AC, Bromwich D, Campbell H, Hamilton GS, Lyons WB, Maasch KA, Aoki S, Xiao C, van Ommen T (2009) State of the Antarctic and Southern Ocean climate system. Rev Geophys 47(RG1003):1–38

Naik SS, Thamban M, Laluraj CM, Redkar BL, Chaturvedi A (2010a) A century of climate variability in the central Dronning Maud Land, East Antarctica and its relation to Southern Annular Mode and El Niño Southern Oscillation. J Geophys Res (Atmospheres) 115:D16102. doi:10.1029/2009JD013268

Naik SS, Thamban M, D'Souza W, Laluraj CM, Rajakumar A, Chaturvedi A (2010b) Influence of climatic teleconnections on the temporal isotopic variability as recorded in a firn core from the central Dronning Maud Land, East Antarctica. J Earth Syst Sci 119:41–49

Nijampurkar VN, Rao DK, Clausen HB, Kaul MK, Chaturvedi A (2002) Records of climatic changes and volcanic events in an ice core from Central Dronning Maudland (East Antarctica) during the past century. Proc Indian Acad Sci (Earth Planet Sciences) 111:39–49

Pickard GL, Emery WJ (1990) Descriptive physical oceanography: an introduction, 5th edn. Pergamon, Oxford, pp 173–176

Schneider DP, Steig EJ, Comiso JC (2004) Recent climate variability in Antarctica from satellite-derived temperature data. J Climate 17:1569–1583

Schneider DP, Steig EJ, van Ommen TD, Bitz CM, Dixon D, Mayewski PA, Jones JM (2006) Antarctic temperatures over the past two centuries, from ice cores. Geophys Res Lett 33: L16707. doi:10.1029/2006GL027057

Steig EJ, Schneider DP, Rutherford SD, Mann ME, Comiso JC, Shindell DT (2009) Warming of the Antarctic ice-sheet surface since the 1957 International Geophysical Year. Nature 457:459–462. doi:10.1038/nature07669

Thamban M, Chaturvedi A, Rajkumar A, Naik SS, D'Souza W, Singh A, Rajan S, Ravindra R (2006) Aerosol perturbations related to volcanic eruptions during the past few centuries as recorded in an ice core from the Central Dronning Maud Land, Antarctica. Curr Sci 91:1200–1207

Thamban M, Laluraj CM, Mahalinganathan K, Redkar BL, Naik SS, Shrivastava PK (2010) Glacio-chemistry of surface snow from the Ingrid Christensen Coast, East Antarctica, and its environmental implications. Antarct Sci 22(4):435–441

Thompson DWJ, Wallace JM (2000) Annular modes in the extratropical circulation. Part I: Month-to-month variability. J Climate 13:1000–1016

Trenberth KE, Caron JM (2001) Estimates of meridional atmosphere and ocean heat transports. J Climate 13:4358–4365

Trenberth KE, Jones PD, Ambenje P, Bojariu R, Easterling D, Klein Tank A, Parker D, Rahimzadeh F, Renwick JA, Rusticucci M, Soden B, Zhai P (2007) Observations: surface and atmospheric climate change. In: Qin SD, Manning M, Chen Z, Marquis M, Averyt KB, Tignor M, Miller HL (eds) Climate change 2007 – the physical science basis. Contribution of WG1 to the fourth assessment report of the Intergovernmental Panel on Climate Change. Cambridge University Press, Cambridge/New York, pp 235–336

Turner J (2004) The El Niño-Southern Oscillation and Antarctica. Int J Climatol 24:1–31. doi:10.1002/joc.965

Turner J, Colwell SR, Marshall GJ, Lachlan-Cope TA, Carleton AM, Jones PD, Lagun V, Reid PA, Iagovkina S (2005) Antarctic climate change during the last 50 years. Int J Climatol 25:279–294

Turner J, Lachlan-Cope TA, Colwell S, Marshall GJ, Connolley WM (2006) Significant warming of the Antarctic winter troposphere. Science 311:1914–1917

Turner J, Bindschadler R, Convey P, di Prisco G, Fahrbach E, Gutt J, Hodgson D, Mayewski P, Summerhayes C (2009) Antarctic climate change and the environment: a contribution to the International Polar Year. Scientific Committee on Antarctic Research, Cambridge. ISBN 978-0-948277-22-1

Van den Broeke MR (2000) The semiannual oscillation and Antarctic climate. Part 5: impact on the annual temperature cycle as derived from NCEP/NCAR re-analysis. Climate Dyn 16:369–377. doi:10.1007/s003820050334

van Ommen TD, Morgan V (1997) Calibrating the ice core paleothermometer using seasonality. J Geophys Res 102(D8):9351–9357

Yuan XJ (2004) ENSO-related impacts on Antarctic sea ice: a synthesis of phenomenon and mechanisms. Antarct Sci 16(4):415–425

Chapter 6
River Systems and River Science in India: Major Drivers and Challenges

R. Sinha, Vikrant Jain, and S.K. Tandon

6.1 Introduction

Rivers, particularly large river systems, constitute one of the most fundamental life-support systems that have sustained civilisations; and are projected to be a critical determinant for the future sustenance of human civilisations. In a world that is expected to witness a near doubling of the human population by the middle of the century, and a world that is likely to be impacted by an unprecedented rate of global change, one of the main sustainability agendas will be the construction of dynamic strategies for the management of natural freshwater systems.

For more than 5,000 years, civilisations have flourished in the South Asian region, leading up to a population hotspot that hosts almost a sixth of the globe's human population. In turn, this has led to significant human interventions and impacts on the freshwater systems of the region. Therefore, the understanding of water problems and water security in this region has to be based on holistic approaches that stress the inter-relationships of earth, water, and humans. As large river systems constitute a lifeline for the future of human populations, it is important to understand them, with the aim of securing their futures and thereby our own futures. In India, amongst other issues, surface runoff and stream flow and discharge patterns of both the Himalayan and the peninsular rivers need detailed rigorous scientific studies. Rigorous analysis of the discharge data of the past few decades of the large river systems of our country are required to build reliable time series that can be used for an improved forecasting of the future discharge trends of these systems.

R. Sinha (✉)
Engineering Geosciences Group, Department of Civil Engineering, IIT, Kanpur, Uttar Pradesh, India
e-mail: rsinha@iitk.ac.in

V. Jain • S.K. Tandon
Department of Geology, University of Delhi, Delhi, India

R. Sinha and R. Ravindra (eds.), *Earth System Processes and Disaster Management*, Society of Earth Scientists Series 1, DOI 10.1007/978-3-642-28845-6_6,

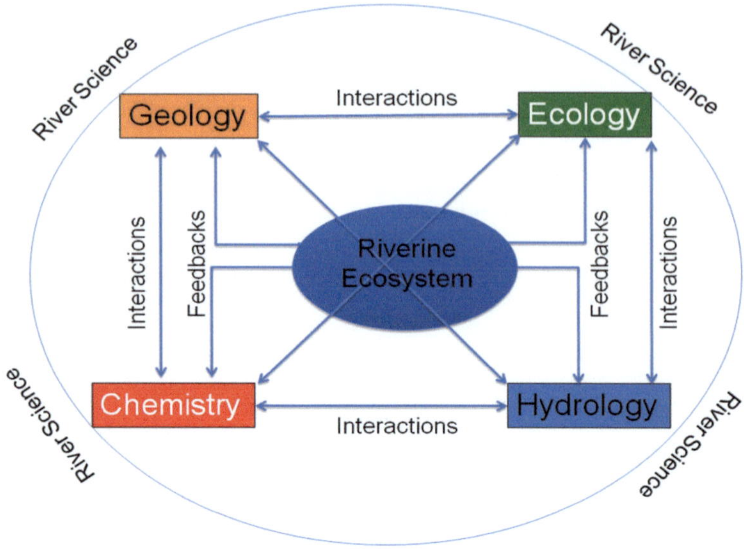

Fig. 6.1 Major components of River Science, interactions and feedbacks

In the future, a growing population will cause the dependence of a greater number of human beings on our river systems. A big question therefore is: can we continue to create 'disturbances' at different scales in the large river systems without obtaining a detailed multi-disciplinary knowledge base of these systems? Human disturbance of different types at different scales in river systems is a consequence of the perceived needs of human populations; however these needs have to be in consonance with the needs of the river itself. For instance, the river requires its 'own space' to perform such functions like flooding and floodplain development. Technological interventions should therefore be practised on the basis of a holistic approach encompassing all aspects of river systems and **River Science** offers such an opportunity.

River Science refers to the study of a variety of processes affecting river systems. This is a truly interdisciplinary science and requires the *explicit joining of two or more areas of understanding into a single conceptual-empirical structure* (Pickett et al. 1994). An all-inclusive definition, provided in a recent report published by the National Academy of Science, is *the study of how hydrological, geological, chemical, and ecological processes interact to influence the form and dynamics of riverine systems and how riverine ecosystems in turn influence these processes across multiple spatial and temporal scales* (NAS 2007). Figure 6.1 illustrates the concept of River Science and shows the interactions among the different disciplines and the feedbacks from the riverine ecosystem. It is important to draw a distinction between 'River Science' and 'Water Science'; while River Science focusses on the processes and fluxes within the channel, Water Science emphasizes water budgeting and resource generation through different storages. River science encompasses all possible

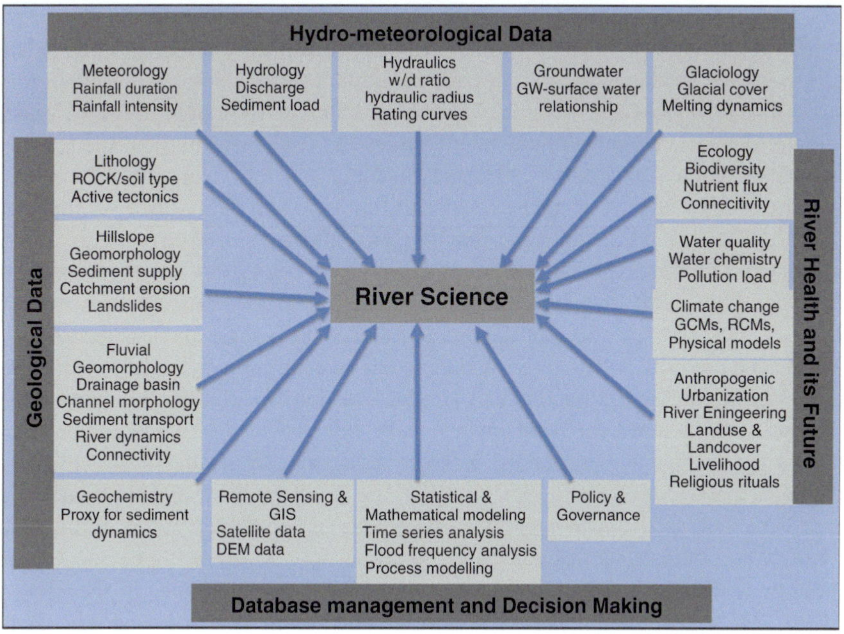

Fig. 6.2 Data required to study River Science; River Science is a multidisciplinary science which requires different set of skills and data to achieve sustainable means for river managements

scales from a single reach to the entire catchment and a variety of data from different disciplines (Fig. 6.2) may be necessary to understand and predict riverine processes.

The NAS (2007) report has identified three topical areas in river science namely, environmental flows and river restoration, sediment transport and geomorphology, and groundwater-surface water interactions which may need special emphasis globally. There is no denying to the fact that human and natural actions have caused the loss or degradation of riverine habitat. The construction of dams and other interventions on several large rivers across the world certainly resulted in short-term benefits, but the large sums of money now being spent on river restoration projects to improve water quality, manage riparian zones, improve habitat, and stabilize streambanks far exceed the benefits accrued. This paper attempts to identify high-priority river science questions in the Indian context, summarizes the present status of data on river science research, and outlines some of the future directions.

6.2 Major Drivers of River Science

Given the importance of rivers as a vital source of freshwater and the ever increasing demands for the same, there has been a pandemic deterioration of river systems globally accompanied by serious threats to biodiversity and water security

(Palmer 2010; Vörösmarty et al. 2010). The very infrastructure projects, which are aimed at the 'development' of water resources, ironically affect the ecological processes and biodiversity. A successful water management strategy must therefore strike a balance between 'water use' for humans and 'ecosystem protection'. Under these circumstances, some of the major drivers and challenges in river science may include ecological restoration, removal of dams and relicensing of hydropower facilities, water allocation, invasive species, climatic variability, urbanization and other land use changes, and water quality (NAS 2007).

In spite of the fact that large sums of money have been spent on river rehabilitation across the globe, the understanding of the science of restoration is fragmentary. In India, most river management programmes have been primarily targeted towards water allocation and water quality and there has been no consolidated effort towards river restoration from an ecological perspective. Further, the exceptionally large expenditures incurred on flood control have not provided sustainable solutions due to the limited understanding of river processes. Even after spending more than Rs. 2,700 crores from 1950 to 1990, the flood problem and flood affected area have increased considerably in India (Agarwal and Narain 1996). While dam removal and relicensing of hydropower facilities are actively being pursued in the western world, such discussions are almost prohibitive in India. In the United States, 241 dams were demolished between 2006 and 2010 (Washington Post, Sep. 7, 2011) which signals a radical shift in river management strategy and marks a new era in river restoration. It has been reported that 85% of dams in the United States will be near the end of their operational lives by 2020 (Washington Post, Sep. 7, 2011). In most cases, the removals have delivered human benefits, e.g. reduction in flooding, as well as ecological ones such as the restoration of fish habitats. Similarly, the Federal Energy Regulatory Commission in the US licenses hydropower dams for 50 years, with possible extensions of 35–50 years. We need a major policy-level intervention in India if a realistic river restoration programme has to succeed.

In the absence of a long-term database on the distribution of key species in long stretches of the river, it is extremely difficult to assess the present ecological status of large rivers such as the Ganga. Basin-scale models of climate impacts have started to appear in the literature (Gosain et al. 2006, 2011; Ghosh and Mujumdar 2008, 2009) but their validation is debatable due to the non-availability of observed hydrological data from the concerned agencies. The assessment of ecological and sociological impacts of climatic variability is even more challenging. Many parts of India are fast getting urbanized and a large-scale change in land use is underway. The quantity and quality of river flows are strongly linked to such changes via runoff generation and a complete understanding of the impacts of such changes is necessary for a sustainable water management, flood risk evaluation, and management of erosion and siltation in river basins. Despite several years of continued efforts and large scale projects, our river water quality remains at risk and several reaches of the most important river in the country, the Ganga, are severely affected by very high pollution loading – both industrial as well as domestic. Any major river rehabilitation project can take a minimum of 25–30 years to allow the

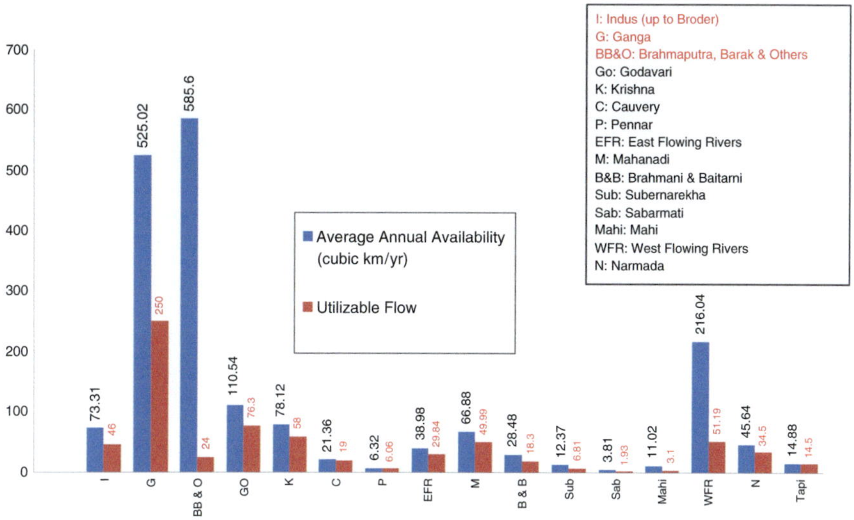

Fig. 6.3 Annual availability and utilizable flow in river basins of India

river to return to a desired state (but not necessarily pristine state), and therefore, it is important that the efforts are sustained and the policy managers are patient.

A scientific approach to tackle such challenges to river management in India requires a highly multi-disciplinary approach, which must be process-based and predictive. The emerging area of river science can provide such an approach.

6.3 Himalayan River Systems – A Major Source of Fresh Water Supply

The Himalaya is host to several large river systems of the world characterized by large catchment size, length, and large volumes of water and sediment discharges (Hovius 1998; Tandon and Sinha 2007). Many of these large rivers drain the Indian subcontinent and have long been the most important source of fresh water supply to the region. Not only is modern society dependent upon rivers for freshwater supply, but ancient civilizations such as the Indus Valley civilization also depended for water resources on them. At present, the total annual flow availability from the rivers draining from the Himalaya as well as Peninsular India is estimated to be ~1,870 km^3/year out of which the Himalayan rivers constitute the major share of ~1,180 km^3/year (Fig. 6.3; Kumar et al. 2005). Utilizable water resource is the quantum of water which can be extracted from its place of natural occurrence and is generally a function of physiographic conditions, socio-political environment, legal and constitutional constraints and the available technology for development. Out of the total available annual flow from rivers, the 'utilizable' flow has been assessed as

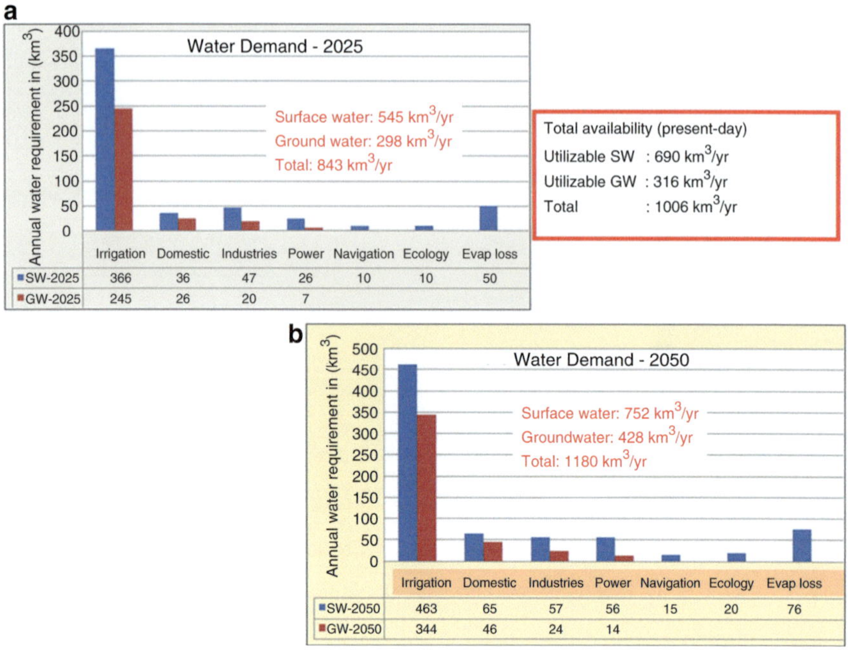

Fig. 6.4 Freshwater demand for future (Data source: Verma and Phansalkar 2007)

~690 km^3/year and again ~50% of this (~320 km^3/year) comes from the Himalayan Rivers (Fig. 6.3). This data clearly demonstrates as to how critical the Himalayan Rivers are in terms of freshwater supply in this region. This also means that the water availability is extremely variable in different regions, for example the regions fed by the Himalayan and peninsular rivers.

The availability of utilizable water in India has been constantly decreasing due to increase in demand for irrigation, industry and domestic use by an ever-increasing population. Figure 6.4a, b show the water demands in various sectors for surface as well as groundwater for the years 2025 and 2050 respectively. The current estimates of total freshwater demand for the years 2025 and 2050 are 843 km^3/year and 1,180 km^3/year respectively (Verma and Phansalkar 2007). Expectedly, the demand for the irrigation sector is the highest (~70%). Comparing this with the presently available resources (1,121 km^3/year including 689 km^3/year for surface water and 432 km^3/year for groundwater), it is clear that India would be close to stretching its limits to meet the demands by the year 2050. However, due to an extremely uneven distribution of freshwater, both spatially as well as temporally, water stress could occur much earlier in some regions.

Freshwater supply from the Himalayan ice-fields and glaciers is a critical resource for a large population in India because they sustain dry-period low flows for major rivers, such as the Indus, Ganges and Brahmaputra Rivers, in the south western Himalaya. The Indus and Ganges Rivers currently have little outflow to the sea during the dry season owing to the large extractions and human interventions. With an ever increasing water demand and projected climate change (discussed later) including the glacier retreat, these outflow volumes could be reduced further. However, the realistic quantitative estimates of hydrological impacts of glacial retreat on the water resources of these rivers are still fragmentary.

Apart from the water shortage and more importantly round-the-year water supply, glacier melting will cause other severe environmental problems. It has been debated that the glacier melt would increase the runoff in the rivers initially but it would cause significant seasonal shifts in water supply, and would increase the flood risks and precipitation variability. Continued and rapid melting of glaciers can lead to the flooding of rivers and to the formation of glacial melt water lakes, which may pose an even more serious threat. Continued melting or calving of ice chunks into lakes can cause catastrophic glacial lake outburst floods such as the one in 1985, in Dig Tsho (Langmoche) Lake in Nepal which killed several people and destroyed important infrastructure facilities and fertile land (Vuichard and Zimmermann 1987).

In addition, increasing flash floods and rockslides, common in upstream Himalayan drainages, degrade roads and trails. Most Himalayan watersheds have experienced a substantial deforestation and overgrazing, making the hillsides much more vulnerable to landslides, either during peak snowmelt, peak monsoon precipitation or in relation to tectonic activity. For an energy constrained economy like India, flow reduction in the major Himalayan Rivers would further limit the possibility of hydropower generation and may have serious consequences about their viability.

6.4 Modern Concepts and Approaches in River Science: The Indian Context

6.4.1 Scale and Hierarchy

Scale is the most important parameter in river studies, as observations, approach and methodology and causality in river studies varies with change in scale. For example, at basin scale, the basin shape and drainage network are important parameters, which are governed by basin geology and climatic conditions (Fig. 6.5). A basin may be consisting of various landscapes, which will be characterized by different set of surface processes. These processes are governed by slope, rainfall, vegetation and riverine fluxes. In a given landscape, river reaches are characterized by valley, floodplain and association of different bars in the channel, which are in turn controlled by variation in stream power and sediment supply. Further, each bar is characterized by variation in grain size in response to

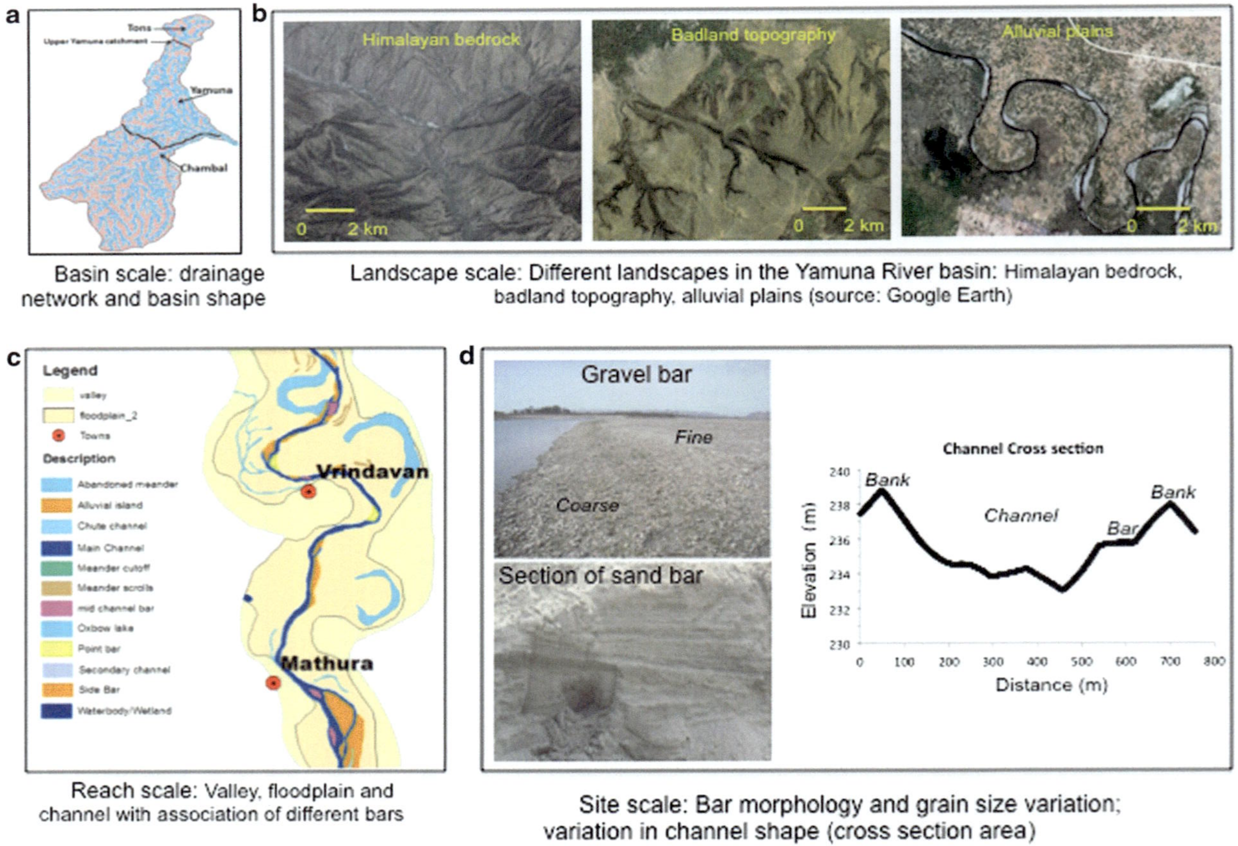

Fig. 6.5 An example of the Yamuna River to highlight different hierarchical levels in a river system. Changes in scale emphasize different characteristics and processes which require different sets of data and causality

site-scale channel hydraulics and processes. Grain size variations and bar morphology are important input parameters for river ecology, as they characterize habitats for different biota (Wilkinson et al. 2006).

These various landforms and geomorphic processes operating at different scales can be linked in a hierarchical model. The hierarchical integration is currently leading to the development of new landscape evolution models in fluvial systems. These landscape evolutionary models have been developed on the basis of knowledge available from analysis of a single grain dynamics in a river system (Allen 2008; Tucker and Hancock 2010). This hierarchical based integration at multiple scales needs to be brought into stream management planning. A good understanding of river processes and morphology is available at different scales; however, these are isolated in nature. Better stream management strategies will require the integration of the available knowledge across the different scales. For example, an important question for stream managers is: how will the deforestation in the Himalaya affect channel morphology, flood hazard and ecological health of a river in downstream plains area? An understanding of this question will require an integration of available knowledge on the landuse-landcover variation and its impact on catchment erosion process (catchment-scale), sediment transport process (site-scale), channel morphological processes (reach-scale), morphology – ecological relationship (reach-scale) and buffering effect of civil engineering structures (site-scale).

6.4.2 *Eco-Geomorphology*

Eco-geomorphology represents an integration of river ecology, hydrology and fluvial geomorphology (Fig. 6.6a) and builds upon eco-hydrology which aimed to bring hydrologists and ecologists together to analyse and predict the flow regime and the ecological status of rivers (Thoms and Parsons 2002). The ecological condition and biotic associations in a river are significantly influenced by the geomorphic condition of the river, and therefore, any efforts towards river rehabilitation must address these issues to derive a long-term benefit. Further, a landform/landscape may respond differently for the same forcing mechanism, and this is governed by landscape/landform sensitivity. *Sensitivity* is defined as the 'propensity for the landscape to undergo a recognizable change in response to changes in the external variables controlling the geomorphic system' (Brunsden and Thornes 1979; Summerfield 1991). The variability of physical environment generally defines the habitat template to which organisms adapt while the temporal and spatial variability of channel processes control both local community composition and adaptive strategies for aquatic and riparian ecosystems (Ward 1989). Geomorphic diversity may be generated in various ways – discharge regimes and vegetation cover (driven by climatic setting and anthropogenic intervention), channel slope and channel form (driven by catchment morphology and tectonics), sediment flux and accommodation (driven by geological settings).

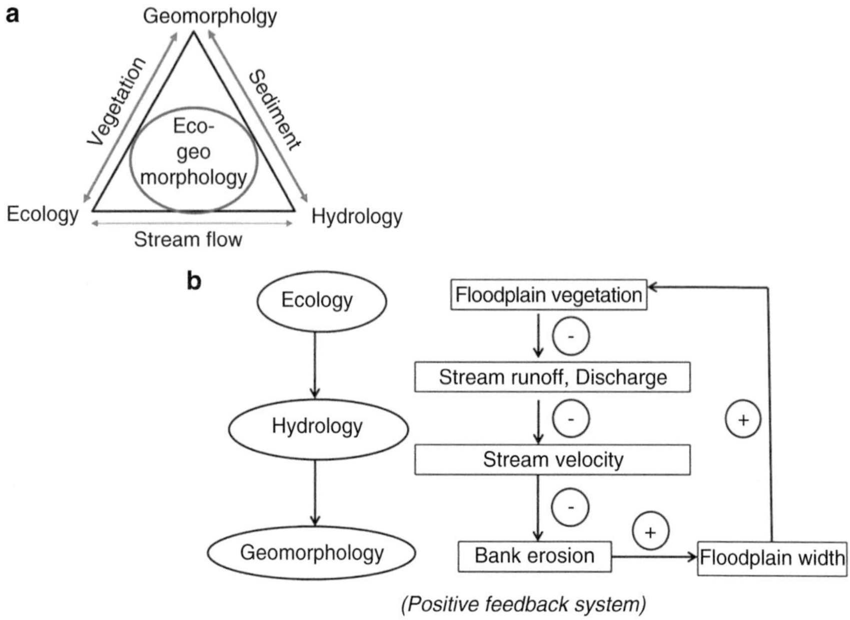

Fig. 6.6 An eco-geomorphological framework for river systems. (**a**) Linkages of ecology, hydrology and geomorphology. (**b**) Eco-geomorphological framework as a positive feedback system

A good example is the relationship between floodplain vegetation, stream runoff and river processes which integrate into a 'positive feedback system' (Fig. 6.6b). Recent research on river systems has also highlighted the importance of understanding controls on channel morphology as a basis for river management and rehabilitation work (Gilvear 1999; Downs and Gregory 2004; Brierley and Fryirs 2005). River morphology not only varies from upstream to downstream in a particular system but also from catchment to catchment in a particular region. Characterization of the geomorphic conditions of river systems provides the basic and first order data set for stream management programs.

As discussed in the previous section, hydrology, ecology and geomorphology are strongly hierarchical disciplines, and therefore, hierarchy becomes the fundamental tenet of eco-geomorphology (Thoms and Parsons 2003). In addition, all these three disciplines are based on several fundamental issues namely scale, connectivity, variability, sensitivity and complexity. Therefore, there is a need to focus on these fundamental issues through an interdisciplinary approach to achieve a scientific framework for the eco-geomorphology of a river system. It is clear that the eco-geomorphological approach needs to be integrative, holistic and process-based with an overarching goal to understand the causal factors of river processes in the appropriate spatial and temporal scales.

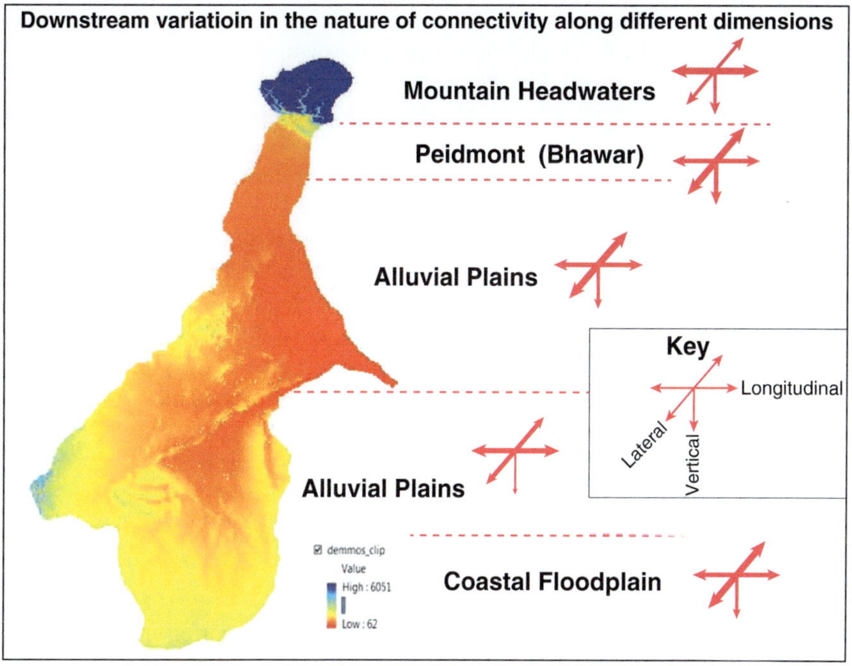

Fig. 6.7 Three dimensions of connectivity and downstream variation in the nature of connectivity. A schematic figure for the Yamuna River is used to highlight downstream variation in connectivity due to change in geomorphic characteristics and processes

6.4.3 Connectivity

A 'river system' necessarily includes the 'morphology' of the river as well as 'connectivity' which is defined as the way in which different landscape compartments fit together in the catchment (Brierley et al. 2006). Connectivity structure in a river system can be studied in four dimensions namely, longitudinal, lateral, vertical and temporal connectivity (Ward 1989; Jain and Tandon 2010). A reach-to-reach connectivity presents an example for longitudinal connectivity. Channel-hillslope or channel-floodplain connectivity is defined as lateral connectivity while groundwater-surface water connectivity represents vertical connectivity. The nature of the connectivity can show significant spatial and temporal variation. A basin may be 'connected' in one dimension but may be 'disconnected' in other dimensions. Further, in a given basin, the nature of connectivity in different dimension varies downstream. An example from a Himalayan river (Yamuna River) shows major types of connectivity in different parts along the channel (Fig. 6.7). While the mountainous headwater reach is characterized by strong longitudinal connectivity due to steep slope, vertical connectivity becomes important in reaches in the piedmont region downstream of the mountain front (*Bhabar zone*) due to high infiltration in coarse-grained sediments. Further downstream, moderately strong lateral and longitudinal connectivity in the plains area are due to higher discharges in river systems. The nature of connectivity will have a significant influence on riverine fluxes, which further governs morphology, habitats and distribution of biota.

Quantification of connectivity is a complex issue and is currently one of the important aspects of geomorphic enquiry. Computation of connectivity index will not only depend on the different dimensions of connectivity but also on the definition of connectivity. Connectivity may be defined through physical connectedness (structural connectivity) or through transfer of material (functional connectivity) (Jain and Tandon 2010). Structural connectivity has been measured through catchment area mapping (Fryirs et al. 2007) or using topographic attributes namely Network Index (Lane et al. 2009). Functional connectivity has been measured using runoff potential and slope characteristics (McHugh et al. 2002) and through storm analysis (Reaney et al. 2007). All these approaches provide quantification of some aspects of connectivity, but more detailed work is required to develop a comprehensive model for connectivity measurement.

6.4.4 Environmental Flow

There is an increasing awareness of allocation of water for human use as well as to continue functioning and maintenance of biodiversity. Water which is allocated and made available for maintaining ecological processes in a desirable state is referred to as the in-stream flow requirement or environmental flows (E-flow). A formal definition of E-flow was coined at the 3rd World Water Forum held in Kyoto in 2003 which says *Environmental Flows is the provision of water within rivers and ground water systems to maintain downstream ecosystems and their benefits, where the river and underground system is subject to competitive uses and flow regulation.* In principle, E-flows are required for maintaining river regimes, making it possible for the river to purify itself, maintaining aquatic biodiversity, performing geomorphic functions, recharging groundwater, and supporting livelihoods. In a country like India where the rivers play a major role in the cultural and the spiritual lives of people, the socio-cultural factors become important considerations in the assessment of E-flow. A range of methodologies for E-flow assessment exist in the literature ranging from purely hydrological (Arthington et al. 2000; King et al. 2003) and purely ecological (Walter et al. 1994) to holistic (King and Louw 1998; King et al. 2000) methods which integrate multiple disciplines (summarized in Tharme 2003).

The assessment of e-flow has gained global importance in recent years keeping in view the rapid increase in extraction of water from rivers. There has been a growing concern about the impacts of the present and future abstractions of water on the ecosystem services involving river systems. There are already initial signals of ecological degradation of river systems across the world (Vörösmarty et al. 2010) and some initial assessment of the Ganga River (WWF 2011) also suggest that several stretches of the river are significantly degraded ecologically. However, the abstraction of water may still be necessary and interventions are unavoidable for meeting human water requirements. In several reaches, it may not be possible to restore the river to its pristine condition. Therefore, the correct perspective of the environmental flow (IUCN 2003) would require the maintenance of a flow pattern (the annual hydrograph) and sediment supply in the rivers in such a manner that the ecosystem services would continue to function in a moderated flow even in a partially degraded state. The issue of sediment flow in rivers has often been ignored in most E-flow assessments and we need to find ways to incorporate sediment flux in the methodology adopted for computing E-flow requirements. In totality, two major parts of the environmental flow can be categorized as (a) biological, to sustain the ecological system, and (b) geomorphological, to maintain the river form and sediment movement necessary for maintaining the basic functions of the river (Bandyopadhyay 2011). It is important to note that the concept of environmental flow does not preclude any human interventions to maintain the desired flow but the challenge is to identify the 'trade offs', which is acceptable to all stakeholders. Bandyopadhyay (2011) advocates for a major policy level intervention in terms of water allocation strategy to prevent the ecosystem services to get restricted, while maintaining both livelihoods and abstraction of water for meeting other economic demands.

A pilot study was carried out by us (RS and VJ) in parts of the Ganga River in India using satellite remote sensing data, DEM data and repetitive field work (WWF 2011). This study supported by WWF-India, involved seven specialist groups namely hydrology, fluvial geomorphology, water quality, hydraulics, biodiversity, livelihoods and cultural-spiritual. Each group identified a set of parameters to characterize the E-flow at a few selected sites. Fluvial geomorphologic studies included the analysis of (a) longitudinal form, (b) cross sectional form and (c) planform of the channel at these sites and derivation of various morphometric parameters. Twelve geomorphic parameters were identified which influence the aquatic life along with the Land use/Land cover in the floodplain, planform dynamics, and channel-floodplain connectivity. These results when integrated with the hydraulic analysis provided estimates of flow depth and flow volumes necessary for channel maintenance from an ecological perspective. Similarly, data from other disciplines were generated in terms of the desired flow and flow motivation forms were prepared. All data were then integrated using the Building Block Methodology (BBM) (King and Louw 1998; King et al. 2000) to generate annual E-flow curves for selected sites for dry season, wet season and flood flows for maintenance and drought years (Fig. 6.8). Although this project followed a rigorous method, the confidence in prediction was limited by validation with the observed data. This work is now being refined as a part of the Ganga River Basin Management Plan (discussed later).

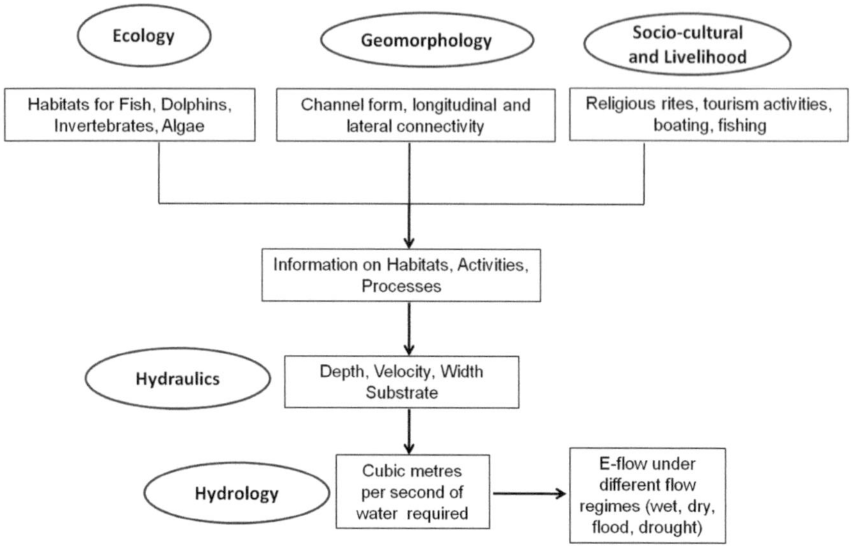

Fig. 6.8 Building Block methodology for E-flow assessment

6.4.5 *River Management and River Health*

River management strategies all around the world have moved from the engineering dominated *command and control* approach to an integrated *ecosystem based approach* that relies on the synthesis of hydrological, geomorphological and ecological data (Downs and Gregory 2003; Brierley and Fryirs 2008; Thorp et al. 2008). Attempts are now on to design engineering solutions using the scientific framework of the river system as the basic template for human intervention. Over the years, the *command and control* approach has been followed in river management, which is a deterministic approach designed for a single purpose and focused on, ‘effects’. Such approaches typically address the problems at site or reach specific scales without serious considerations of upstream and downstream consequences and related connectivity issues. Even though provisions for ecosystem protection were made, they were minor issues and were never implemented fully. On the contrary, the *ecosystem based* approach is a cross-disciplinary, holistic approach, focused on ‘causes’ and applied at catchment to reach scale. This is a probabilistic approach that recognizes uncertainty and complexity in the system. The physical template of a river system provides the basic structure to analyze the different aspects in an integrated approach.

The human relationship to any given river system is a key factor for ensuring healthier river futures, and the importance of place in designing rehabilitation initiatives cannot therefore be underestimated. In the past, humans have made interventions at different scales in river systems in order to use them. These interventions are largely through river engineering which is a discipline that stresses the utilitarian aspects of river systems rather than their evolutionary and their multi-disciplinary aspects. It is now increasingly being realized that river engineering should be practiced on a platform of River Science.

Most large river systems provide critical ecosystem services to large populations in South Asia. As a consequence, considerable pressure is exerted by these populations on the river systems, thereby impacting the health condition of these river systems. The scales of the anthropogenic disturbances on the river systems are variable, and principally the large river systems have been impacted by high quantities of water abstraction for irrigation, pollution by industrial and urban expansion, as well as reduction and loss of connectivity because of the construction of dams, barrages, and related engineering structures.

6.5 Impacts of Future Climate Change and Challenges Ahead

In addition to the increased demand due to growing population, the freshwater supply in future will also be influenced by climate change impacts such as monsoon variability (Goswami et al. 2006; Mall et al. 2006; Immerzeel et al. 2010) and retreating glaciers (Raina 2006; Kulkarni et al. 2007, 2010; Raina and Srivastava 2008). Analysis of the daily rainfall for the last five decades has shown a significant rising trend in frequency and magnitude of extreme rain events although the monsoon system as a whole has remained stable (Goswami et al. 2006). Although there is some data now on the trends of rainfall in some river basins of India (Sontakke and Singh 2008; Kumar and Jain 2010) there are still large uncertainties in the future trends of rainfall and the assessment of hydrologic impacts. The available models suggest that there may be an increase in precipitation in majority of the river systems for the Greenhouse Gas (GHG) scenario (Gosain et al. 2006, 2011). However, the corresponding runoff for these basins may not necessarily increase due to increases in evapo-transpiration on account of corresponding increased temperatures. The Ganga system is a good example to illustrate this point where a 2.5% increase in precipitation may be accompanied by ~7% increase in evapo-transpiration by 2030s (Gosain et al. 2011). The results published by Gosain et al. (2011) suggest that most large river basins except the Brahmaputra and the Cauvery will have an increased precipitation at the basin level in a GHG scenario and there will be an associated increase in water yield in these basins for the next 20 years or so. However, the situation changes drastically when the modelling period changes to another 50 years (2080s) under which ET increases by 40% for a majority of the basins due to increase in temperature and enhanced precipitation. However, the authors note a significant spatial variability in these large basins when the models were run for different sub-basins. However, these models were run for virgin flow conditions (without incorporating the effects of human interventions such as dams and barrages) and did not consider any land use

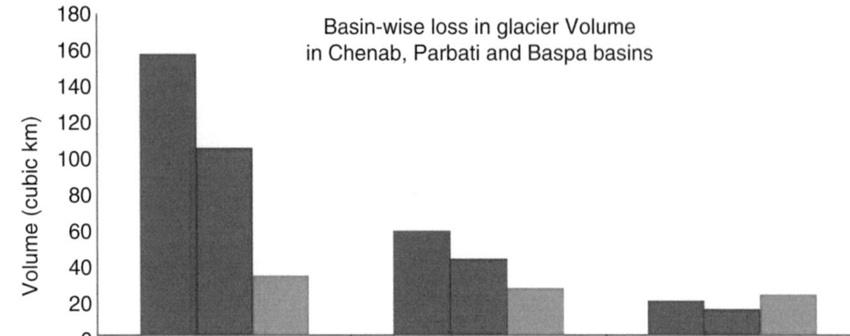

	Chenab	Parbati	Baspa
1962	157.6	58.5	19.1
2001-04	105.03	43	14.7
loss(%)	33.33	26.5	23

Fig. 6.9 Basin-wise loss in glacier volume in Chenab, Parbati and Basapa basins (Data from Kulkarni et al. 2007)

changes over time. Moreover, the hydrological impacts on geomorphology and ecosystem services are yet to be assessed.

The Himalayan glaciers that feed several of the great rivers of India (e.g. the Ganga, Indus, Brahmaputra) and ensure a year-round water supply to these river systems have been retreating at several places although the rates of retreat are variable and debatable (Kulkarni et al. 2007, 2010, Raina, this volume). Data on the loss of glacier area in the Himalaya and the resulting loss in river flows are highly debated, but it is extremely variable spatially. The controversial IPCC (2007) report on Himalayan glacier suggesting that the possible disappearance by 2035 has been discredited by many workers as it would require an increase of melt rates by about 20 times the current observed melt rate in the Himalaya (USAID 2010). Some published data for the Chenab, Pārbati and Basapa basins (Kulkarni et al. 2007) (Fig. 6.9) suggest that the percentage loss of glacier volume ranges from 23% to 33.3% during the period 1962–2004. In a more recent publication, the glacial retreat for 1,868 glaciers in 11 basins distributed across the Indian Himalaya for the period 1962–2001/2004 was estimated (Kulkarni 2010). These estimates show that the overall glacial area has decreased from 6,332 to 5,329 km^2 which amounts to an average loss of ~16%. In the Nepal Himalaya, the glaciers are reported to be retreating at rates of 10–60 m per year and many small glaciers (<0.2 km^2) have already disappeared (Bajracharya et al. 2007). In the eastern Himalaya around Kathmandu, Kayastha and Harrison (2008) noted no significant trend in the precipitation records up to the year 2000. In contrast, in the western Himalaya, several examples of decreasing rates of retreat are reported, such as those from observations on the Gangotri Glacier (Kumar et al. 2008) and in the Karakoram (Hewitt 2005). Mass balance computation of the Siachen Glacier in the Nubra Valley, eastern Karakoram, India for the period 1986–1991 showed a negative trend except for the period 1988–1989 when a positive mass balance (+0.35 m) was attributed to comparatively heavy winter snowfall amounts and low temperatures during the ablation season (Bhutiyani 1999).

Early estimates of the loss of glacier melt water contribution for the Ganga basin suggested that the July–September flows would reduce by two thirds, causing water shortages for 500 million people and 37% of India's irrigated land (Jain 2001; Singh et al. 1994). Similarly, very high estimates of ~70% reduction in the flow of the Ganga river in summer months (Singh and Bengtsson 2004; Barnett et al. 2005) are probably unrealistic. However, several other workers have estimated a decrease of 10–20% in the summer flow of the Himalayan Rivers depending upon their geographical location (Alford 1992; Rees and Collins 2006). The effects are expected to be more severe in the western part in comparison to the eastern part in line with the present-day rainfall distribution in the Himalayan foreland basin. More recent modeling effort by Immerzeel et al. (2010) across the Greater Himalayan region has used Normalized Melt Index (NMI) defined as the volumetric snow and glacier upstream discharge divided by the downstream natural discharge, to estimate the role of melt water in river discharges. Results from the NMI analysis suggest that the glacier melt water is extremely important in the Indus Basin (NMI = 1.5) whereas it is reasonably important for the Brahmaputra River (NMI = 0.27) but only plays a modest role for the Ganga River (NMI = 0.1). It is clear therefore that the accurate assessment of glacier retreat and the role of glacier melt on river hydrology are largely unaccomplished. Moreover, the initial assessment of glacier melt vulnerability has highlighted the human health and ecosystem services are likely to be impacted from the changes in the hydrologic system (USAID 2010).

6.6 River Science in the Indian Context – Priority Areas and Major Research Questions

The Indian sub-continent has several large river systems, defined generally on the basis of physical parameters, namely channel length (>2,500 km) or basin area (800,000 km^2) (Tandon and Sinha 2007). A large river system is a hierarchical entity, which is characterized by complex and nonlinear response due to different stages of threshold, connectivity and equilibrium conditions in its smaller compartments. Therefore, the prediction of the future geomorphic state of any large river system is one of the most challenging research tasks, especially in the scenario of climate change threat and increasing human interferences. The priority areas of river science in India may include assessment and (re)allocation of water resources, eco-geomorphology, environmental flow, and river rehabilitation, river hazards such as river dynamics and floods, and hydrologic response to climate change including extreme events.

As discussed earlier, the projected future water demands already exceed the water availability from surface and groundwater resources. Therefore, the only way to meet the future demands would be to relook into the water allocation for different sectors such as irrigation, domestic, industrial etc. and to encourage more efficient water management practices. Such (re)allocation would also have to take into account climate change scenarios. In a recent paper, it has already been pointed out that improved irrigation efficiency would prove to be an extremely effective way to mitigate the impacts of climate change on water resources (Gosain et al. 2011). The present water management practices in India do not take into account ecosystem services in an appropriate manner, and therefore, the understanding of eco-geomorphology would provide important inputs for ecosystem-based management of rivers. The assessment of environmental flow needs to be done urgently for most rivers across the country and major efforts should be directed towards the selection of an appropriate methodology, which should be holistic and targeted to cater to all stakeholders. Once we have an assessment of the present ecological state of the river and the environmental flow requirements, the strategies for river restoration will have to be designed so that the ecological state of the rivers can be improved to a desired level while allowing the abstractions for human requirements.

While several rivers are ecologically degrading due to large scale abstractions and reduction in flow, the rivers such as the Kosi and its tributaries, and the Gandak in north Bihar and the Brahmaputra in Assam are notorious for their dynamic behavior, repeated flooding and extensive bank erosion (Wells and Dorr 1987; Jain and Sinha 2004; Sarma 2005; Mishra 2008; Sinha 2009; Chakraborty et al. 2010). In many regions such as the Kosi, the flood magnitude and frequency have increased due to human interventions (Sinha 2009) and there is a need to re-evaluate the efficacy of these interventions. Finally, the data presented in the previous section clearly demonstrates that the impact of climate change on our river systems is extremely uncertain and there is an urgent need to develop coupled RCM-hydrological models to assess the impact of climate change on water availability and agricultural productivity.

Some of the specific research questions in river science in the Indian context may be framed as follows:

1. How will climate change affect the *hydrological budget* of the river system, for example, contribution from glacial melt and precipitation, and how will this change impact the Himalayan Rivers, their geomorphology, sediment transport characteristics and connectivity?
2. How can high-resolution (10^2–10^3 years) proxy based climate data or climate modeling results (downscaling of GCMs etc.) be integrated with stream hydrology and fluvial processes?
3. How is the river morphology related to current discharge and sediment load and how it is responding to changes in hydrologic regime due to natural and anthropogenic processes? How to differentiate the impacts of natural and anthropogenic impacts on river systems?

4. What is the sensitivity of the different reaches to known external forcings and how can such differential sensitivity help to prioritize management planning in a large river?
5. What is the pattern of energy distribution and sediment supply in the trunk channel and its tributaries? What are the inherent geological and geomorphological controls and how do they affect this distribution?
6. What is the connectivity structure and what is its spatio-temporal dynamics for large river system? How are patterns of energy and flux distribution related with the connectivity structure?
7. How can we model the future geomorphological (physical) changes on the basis of modeled future changes in flux and energy, i.e. prediction of geomorphic change on the basis of climate change prediction model?
8. What is the relationship between geomorphology and biodiversity? How will the future changes in geomorphological setting affect the riverine ecosystem and biodiversity?
9. How have urbanized centers (megacities) affected the physical–chemical characteristics of the reaches of a large river and the river as a whole?
10. What is the nature of river–groundwater connectivity and what are its implications on water resources management and riverine ecosystem management?
11. What are the limits to human disturbance of a large river system? How do we define and estimate the values of environmental-flow at different reaches?

It is obvious that these research questions would require a multi-disciplinary approach involving geomorphologists, engineers, ecologists and social scientists. River Science provides an ideal platform to tackle such issues and it is time now to promote this science in India in a major way.

6.7 Recent Initiatives – Ganga River Basin Management Plan (GRBMP)

A large programme on Ganga River Basin Management Plan (GRBMP) has been initiated by the Ministry of Environment and Forests, Government of India which aims to provide a basic framework for developing a river management plan based on five fundamental premises: (1) river must continuously flow, (2) river must have longitudinal and lateral connectivity, (3) river must have adequate space for its various functions, (4) river must function as an ecological entity, and (5) river must be kept free from any kind of wastes. Indian Institute of Technology Kanpur is coordinating this programme with participation from other IITs and several universities and institutions across the country. The Ganga River Basin is a multi-faceted system and requires multidisciplinary and interdisciplinary approach. For the integrated management of the Ganga River Basin several aspects need to be considered. Therefore, this programme is being undertaken through several thematic groups viz. Environmental Quality and Pollution (EQP), Water Resources Management (WRM), Fluvial Geomorphology (FGM), Ecology and Biodiversity (ENB), Socio-Economic and Cultural (SEC), Policy, Law and Governance (PLG), Geo-spatial Data Base Management (GDM) and Communication (COM). The teams working on each of the themes are closely interacting and it is expected that a comprehensive plan will emerge.

This initiative emphasizes on the eco-geomorphologic approach for river management and is the first of its kind to be implemented on a large river system such as the Ganga. While variability of the physical environment generally defines the habitat template to which organisms adapt, the temporal and spatial variability of geomorphic processes have particularly important controls on both local community composition and adaptive strategies for aquatic and riparian ecosystems. Keeping this in view, the geomorphic component of this programme has been designed to (a) prepare and compile geomorphic map of the Ganga River and classification of the reaches in terms of their geomorphic condition, (b) map the patterns of river dynamics at different reaches and to understand the causative factors, (c) generate stream power distribution pattern of various reaches of the Ganga river and analysis of its variation in the Ganga River (after Jain et al. 2006), (d) determine the effects of river energy and sediment supply as controls on channel morphology, (e) assess the hydrological-geomorphological-ecological relationships to develop tool for monitoring river health and sustainable river management based on River Styles Framework, and (f) define environmental flow for different reaches on the basis of geomorphic conditions.

We are currently using River Styles® Framework (Brierley and Fryirs 2005) to integrate diverse parameters for river management which involves four stages of investigation. The first stage focuses on *identification, interpretation and mapping* of River Styles throughout the river catchment. The second stage involves assessing the geomorphic condition of each reach of each River Style in the catchment. By placing each of these reaches in their catchment context, along with an interpretation of limiting factors, the geomorphic recovery potential of the reaches of each River Styles will be determined. From this, predictions of likely future condition will be determined in the third stage of investigation. Finally, with this information in hand, realistic target conditions for river rehabilitation programs can be identified for each reach, framed within a catchment-based vision. Working with local/regional river managers, a physically-meaningful framework for management strategies for river rehabilitation and conservation would then be applied.

Meaningful river management strategies must build on explanation of what a river looks like, how it behaves, and how it is likely to adjust in response to changing linkages of physical processes within a catchment (Gilvear 1999; Gregory 2003). The hydrology-geomorphology-ecology relationship for the Ganga River will be the first such initial tool which will not only explain the geomorphic variability in the river system, but will also help to quantify the required hydrological changes to achieve desired geomorphic condition of the river. The project will provide a process-based understanding of the present geomorphic condition of the river. It will also help to characterise, explain and predict the future changes in the geomorphic condition of the river. The understanding will act as the basic template to carry out sustainable stream management programmes.

However, geomorphic changes are slow processes and in places they may be irreversible if they have crossed a threshold condition. Therefore, (1) *the determination of river recovery potential* and (2) *river recovery path* along with (3) *quantification of threshold condition of geomorphic change in a river* will be required to have quantitative estimate of geomorphic change in the river condition. Further, the determination of recovery potential will need the determination of *three-dimensional connectivity* in the basin area.

6.8 Concluding Remarks

One of the tenets of River Science is to respect the inherent diversity, complexity, and variability of river systems. This however cannot be done in a reductionist discipline-specific mode. Looking at the uncertain availability of water and the escalating future demands, it would be pertinent to consider fresh water as a 'limited resource'. Considering the role of Himalayan river systems as a critical source of freshwater supply, it is necessary to adopt a sustainable river management strategy. More than the issue of 'demand and supply', there is now a strong realization and effort to maintain the 'river health' and 'environmental flow' which are significantly influenced by geomorphic characteristics and biotic association of the river. Anthropogenic modifications on the Himalayan river systems, mostly driven by increased demands for fresh water, have been extensive and any efforts towards maintaining sustainable flow and river rehabilitation must address these issues to derive a long-term benefit.

Acknowledgements This is a contribution to IGCP 582 on Tropical Rivers.

References

Agarwal A, Narain S (1996) Floods, floodplains and environmental myths. State of India's Environment: a citizen report. Centre for Science and Environment, New Delhi

Alford D (1992) Hydrological aspects of the Himalayan region, ICIMOD occasional paper no. 18, Kathmandu, p 68

Allen PA (2008) From landscapes into geological history. Nature 451:274–276

Arthington AH, Brizga SO, Choy SC, Kennard MJ, Mackay SJ, McCosker RO, Ruffini JL, Zalucki JM (2000) Environmental flow requirements of the Brisbane river downstream of Wivenhoe dam. South East Queensland Water Corporation, and Centre for Catchment and In-Stream Research, Griffith University, Brisbane

Bajracharya SR, Mool PK, Shrestha BR (2007) Impacts of climate change on Himalayan glaciers and glacial lakes. International Center for Integrated Mountain Development (ICIMOD) and UNEP, Kathmandu, p 119

Bandyopadhyay J (2011) Deciphering environmental flows. Seminar (Online) 626:50–55

Barnett TP, Adam JC, Lettenmaier DP (2005) Potential impacts of warming climate on water availability in snow-dominated regions. Nature 438:303–309
Bhutiyani MR (1999) Mass-balance studies on Siachen Glacier in the Nubra valley, Karakoram Himalaya, India. J Glaciol 45(149):112–118
Brierley GJ, Fryirs KA (2005) Geomorphology and river management: applications of the River Styles Framework. Blackwell, Oxford, p 398
Brierley GJ, Fryirs KA (eds) (2008) River futures: an integrative scientific approach to river repair. Island Press, Washington, DC
Brierley GJ, Fryirs K, Jain V (2006) Landscape connectivity: the geographic basis of geomorphic applications. Area 38:165–174
Brunsden D, Thornes JB (1979) Landscape sensitivity and change. Trans Inst Br Geogr NS 4:303–484
Chakraborty T, Kar R, Ghosh P, Basu S (2010) Kosi megafan: historical records, geomorphology and the recent avulsion of the Kosi river. Quat Int 227:143–160
IPCC [Intergovernmental Panel on Climate Change] (2007) Climate change 2007: synthesis report. Contribution of working groups I, II and III to the fourth assessment report of the intergovernmental panel on climate change [Core Writing Team, Pachauri RK, Reisinger A (eds)]. IPCC, Geneva
Downs P, Gregory K (2004) River channel management: towards sustainable catchment hydrosystems. Hodder Arnold, London, p 395
Fryirs K, Brierley GJ, Preston NJ, Spencer J (2007) Catchment-scale (dis)connectivity in sediment flux in the upper Hunter catchment, New South Wales, Australia. Geomorphology 84:297–316
Ghosh S, Mujumdar P (2008) Statistical downscaling of GCM simulations to streamflow using relevance vector machine. Adv Water Resour 31:132–146
Ghosh S, Mujumdar P (2009) Climate change impact assessment – uncertainty modeling with imprecise probability. J Geophys Res Atm 114:D18113
Gilvear DJ (1999) Fluvial geomorphology and river engineering: future roles utilizing a fluvial hydrosystem framework. Geomorphology 31:229–245
Gosain AK, Rao S, Basuray D (2006) Climate change impact assessment on hydrology of Indian river basins. Curr Sci 90:346–353
Gosain AK, Rao S, Arora A (2011) Climate change impact assessment of water resources of India. Curr Sci 101:356–371
Goswami BN, Venugopal V, Sengupta D, Madhusoodanan MS, Xavier PK (2006) Increasing trend of extreme rain events over India in a warming environment. Science 314:1442–1445
Gregory KJ (2003) Palaeohydrology, environmental change and river channel management. In: Gregory KJ, Benito G (eds) Palaeohydrology, understanding global change. Wiley, Chichester, pp 357–378
Hewitt K (2005) The Karakoram anomaly? Glacier expansion and the 'elevation effect,' Karakoram, Himalaya. Mountain Research and Development 25(4):332–340
Hovius N (1998) Controls on sediment supply by large rivers. In: Shanley KW, McCabe PJ (eds) Relative role of eustasy, climate and tectonism in continental rocks. SEPM Special Publication No. 59, Tulsa, pp 4–15
Immerzeel WW, van Beek LPH, Bierkens MFP (2010) Will climate change affect the Asian water towers? Science 238:1382–1385
IUCN, Flow (2003) The essentials of environmental flows. IUCN, Gland
Jain CK (2001) A hydro-chemical study of a mountainous watershed: the Ganga, India. Water Res 36(5):1262–1274
Jain V, Sinha R (2004) Fluvial dynamics of an anabranching river system in Himalayan foreland basin, north Bihar Plains, India. Geomorphology 60(1–2):147–170
Jain V, Tandon SK (2010) Conceptual assessment of (dis)connectivity and its application to the Ganga river dispersal system. Geomorphology 118:349–358

Jain V, Preston N, Fryirs K, Brierley G (2006) Comparative assessment of three approaches for deriving stream power plots along long profiles in the upper Hunter River catchment, New South Wales, Australia. Geomorphology 74(1–4):297–317

Kayastha RB, Harrison SP (2008) Changes of the equilibrium-line altitude since the Little Ice Age in the Nepalese Himalaya. Ann Glaciol 48:93–99

King JM, Louw MD (1998) Instream flow assessments for regulated rivers in South Africa using the Building Block Methodology. Aquat Ecosyst Health Manage 1:109–124

King JM, Tharme RE, De Villiers M (eds) (2000) Environmental flow assessments for rivers: manual for the building block methodology. Water Research Commission Technology Transfer Report No. TT131/00. Water Research Commission, Pretoria

King J, Brown C, Sabet H (2003) A scenario-based holistic approach to environmental flow assessments for rivers. River Res Appl 19:619–639

Kulkarni AV (2010) Monitoring Himalayan cryosphere using remote sensing techniques. J Indian Inst Sci 90(4):457–469

Kulkarni AV, Bahuguna IM, Rathore BP, Singh SK, Randhawa SS, Sood RK, Dhar S (2007) Glacial retreat in Himalaya using Indian remote sensing satellite data. Curr Sci 92(1):69–74

Kumar V, Jain S (2010) Rainfall trends in Ganga-Brahmaputra-Meghna river basins of India (1951–2004). Hydrol J 33(Special Issue):59–66

Kumar R, Singh RD, Sharma KD (2005) Water resources of India. Curr Sci 89(5):795–811

Kumar K, Dumka RK, Miral MS, Satyal GS, Pant M (2008) Estimation of the retreat of Gangotri glacier using rapid static and kinematic GPS survey. Curr Sci 94(2):258–262

Lane SN, Reaney SM, Heathwaite AL (2009) Representation of landscape hydrological connectivity using a topographically driven surface flow index. Water Resour Res 45:W08423. doi:10.1029/2008WR007336

Mall RK, Gupta A, Singh R, Singh RS, Rathore LS (2006) Water resources and climate change: an Indian perspective. Curr Sci 90(12):1610–1626

McHugh M, Harrod TR, Morgan RPC (2002) The extent of soil erosion in upland England and Wales. Earth Surf Process Landform 27(1):99–107

Mishra DK (2008) The Kosi and the Embankment Story, Economic and Political Review, V.43 (46), Nov. 2008

NAS (2007) River Science at the United States Geological Survey, 2010. National Academic Press, Washington, DC, p 206

Palmer MA (2010) Water resources: beyond infrastructure. Nature 467:534–535

Pickett STA, Kolasa J, Jones CG (1994) Ecological understanding: the nature of theory and the theory of nature. Academic, Orlando, p 206

Raina VK (2006) Glacier retreat and global warming – a review. XIX Indian Institute of Geomorphologists National Conference on Himalaya to Indian Ocean, geomorphic processes and landscape change, University of Jammu

Raina VK, Srivastava D (2008) Glacier atlas of India. Geological Society of India, Bangalore

Reaney SM, Bracken LJ, Kirkby MJ (2007) Use of the connectivity of runoff model (CRUM) to investigate the influence of storm characteristics on runoff generation and connectivity in semi-arid areas. Hydrol Process 21:894–906

Rees HG, Collins DN (2006) Regional differences in response of flow in glacier-fed Himalayan rivers to climatic warming. Hydrol Process 20(10):2157–2169

Sarma JN (2005) Fluvial processes and morphology of the Brahmaputra River in Assam, India. Geomorphology 70(3–4):226–256

Singh P, Bengtsson L (2004) Impact of warmer climate on melt and evaporation for the rain-fed, snow-fed and glacier-fed basins in the Himalayan region. J Hydrol 300:140–155

Singh PS, Jain SK, Kumar N, Singh UK (1994) Snow and glacier contribution in the Ganga river at Deoprayag. Technical report CS (AR)132. National Institute of Hydrology, Roorkee, pp 1993–1994

Sinha R (2009) The great avulsion of Kosi on 18 August 2008. Curr Sci 97(3):429–433

Sontakke NA, Singh N (2008) Instrumental period rainfall series of the Indian region (1813–2005): revised reconstruction, update and analysis. Holocene 17:1055–1066

Summerfield SA (1991) Global geomorphology: an introduction to the study of landforms. Longman/Wiley, London/New York, p 537

Tandon SK, Sinha R (2007) Geology of large river systems, Chap 2. In: Gupta A (ed) Large rivers. Wiley, Chichester/England

Tharme RE (2003) A global perspective on environmental flow assessment: emerging trends in the development and application of environmental flow methodologies for rivers. River Res Appl 19:397–441

Thoms MC, Parsons M. (2002) Eco-geomorphology: an interdisciplinary approach to river science. In: The structure, function and management implications of fluvial sedimentary systems (Proceedings of an international symposium held at Alice Springs, Australia). IAHS Publication No. 276, pp 113–119

Thoms MC, Parsons ME (2003) Identifying spatial and temporal patterns in the character of the Condamine Balonne river, Australia, using multivariate statistics. River Res Appl 19:443–458

Thorp JH, Thomas MC, Delong MD (2008) The riverine ecosystem synthesis: towards conceptual cohesiveness in river science. Elsevier, Amsterdam, p 208

Tucker GE, Hancock GR (2010) Modelling landscape evolution. Earth Surf Process Landform 35(1):28–50

USAID (2010) Changing glaciers and hydrology in Asia addressing vulnerabilities to glacier melt impacts, Nov 2010

Verma S, Phansalkar SJ (2007) India's water future 2050: potential deviations from 'Business-as-Usual'. Int J Rural Manage 3(1):149–179

Vörösmarty CJ, McIntyre PB, Gessner MO, Dudgeon D, Prusevich A, Green P, Glidden S, Bunn SE, Sullivan CA, Liermann CR, Davies PM (2010) Global threats to human water security and river biodiversity. Nature 467:555–561

Vuichard M, Zimmermann M (1987) The 1985 catastrophic drainage of a moraine-dammed lake, Khumbu Himal, Nepal: cause and consequences. Mount Res Dev 7(2):91–110

Walter AC, Burgess GK, Johnston PJ (1994) Assessment of a process for determining environmental flows. In: Environmental flows seminar proceedings. AWWA Inc. Artarmon, pp 195–201

Ward JV (1989) The four-dimensional nature of lotic ecosystems. J N Am Benthol Soc 8(1):2–8

Wells NA, Dorr JA (1987) Shifting of the Kosi river, northern India. Geology 15:204–207

Wilkinson MJ, Marshall LG, Lundberg JG (2006) River behavior on megafans and potential influences on diversification and distribution of aquatic organisms. J S Am Earth Sci 21:151–172

WWF (2011) Assessment of environmental flows for the Upper Ganga basin. WWF-India Secretariat, Lodhi Estate, New Delhi

Section II
Earth's Climate System and Paleoclimate

Chapter 7
Impact of Climate Change on Lichen and Moss Communities in Ny-Ålesund, Arctic: Some Preliminary Observations

S.M. Singh and Rasik Ravindra

7.1 Introduction

Ny-Ålesund is on the west coast of Spitsbergen, the largest island of Svalbard archipelago located ~1,200 km from the North Pole. Topographical features of Ny-Ålesund include eastern and western glaciers, terminal moraines, glacial streams and rivers flowing northwards to Kongsfjord. Within the marine terraces, gravelly and stony plains are found to be dominant. River system is composed of numerous streams swollen with melted snow, which flow in gullies, rock-crevices and under high rock strike ridges. Intensive snow melting leads to formation of numerous small lakes that disappear later. Carboniferous, Triassic and younger period rocks are seen exposed in the area. Climatic changes in this region are occurring at a faster pace than anywhere else on the planet. Though, climatic conditions are unfavorable for the development of plants, short vegetative period (2–3 months) and low air temperature render the development of many species. Two seasons are distinguished in the region: a long winter and a very short summer. The mean temperature in the coldest month (February) is $-14°C$ while the warmest month (July) has a mean temperature of $+5°C$.

Lichen and mosses play a predominant role in the terrestrial ecosystems of Ny-Ålesund. They have a unique survival and functioning mechanism in the extremities of environment due to their novel biochemical adaptations. They constitute one of the main elements of tundra along with algae and angiosperms. The lichenological investigation in Svalbard, Arctic was initiated by Lynge (1938), Mattick (1949), Hafellner (1982), Elvebakk (1984), Olech (1987), Olech and Alstrup (1989), Søchting and Olech (1995) etc. Elvebakk and Hertel (1996) reviewed entire Svalbard lichens and published a checklist of 597 species from the area. Bryophytes

S.M. Singh (✉) • R. Ravindra
National Centre for Antarctic and Ocean Research, Head Land Sada, Vasco da Gama, Goa 403 804, India
e-mail: smsingh@ncaor.org

R. Sinha and R. Ravindra (eds.), *Earth System Processes and Disaster Management*, Society of Earth Scientists Series 1, DOI 10.1007/978-3-642-28845-6_7,

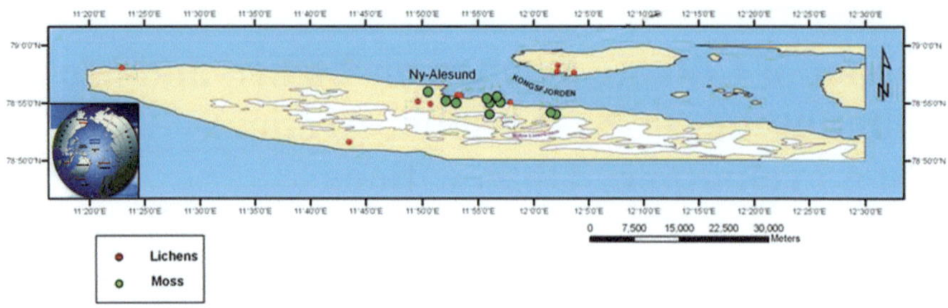

Fig. 7.1 Distribution of lichen and moss communities in the area of Ny-Ålesund

were mentioned from Svalbard, Arctic for the first time by Martens (1675). Later Phips (1777), Dillenius (1741), Lindberg (1883), Kuc (1973); Frisvoll and Elvebakk (1996) etc. reported 288 mosses from the area. Recently, a new species *Meesia hexasticha* as the northernmost record from Ny-Ålesund has been added to the total moss flora of Spitsbergen (Ochyra et al. 2009).

The information on community structure of lichens and mosses, and their distribution in the area of Ny-Ålesund Arctic is fragmentary. Keeping this view in mind, studies on lichen and mosses of Ny-Ålesund, Arctic were started. In the present study, a survey was conducted and several rocks and soil samples containing lichens and mosses were collected during Indian Arctic Expedition 2007 (Fig. 7.1). The identification of samples from different communities of Ny-Ålesund resulted in 34 lichen and 12 moss species (Table 7.1a, b). The identified specimens are archived in herbarium of National Centre for Antarctic and Ocean Research, Goa. The literature of Thompson (1984, 1997) and Olech and Alstrup (1989) were followed for the identification of species.

7.2 Community Structure

It was observed that the lichen flora of the Ny-Ålesund is comprised of foliose, fruticose, crustose and squamulose lichens. *Cetraria fastigiata,* a widely distributed lichen is the most dominant species in the island followed by *Cetraria delisei, Flavocetraria cucullata, Cetraria ericetorum, Physconia muscigena* and *Umbilicaria* species. Rock is the major substratum in the island accommodating lichens followed by moss species. Tundra lichens appear in a variety of colors such as orange, black, brown, cream and pale green. True soil was virtually rare in the study area, but a thin layer of soil probably accumulated in the rock crevices, base of the rock, in moss beds and in plain grounds. The closely packed soil grain forms a hard crust, a suitable habitat for lichens and mosses. The dominating types of soil are primitive lithosols and regosols. Only a small area was covered with plants characteristic of the tundra (Fig. 7.2a). There were four prominent areas bearing the

Table 7.1 List of lichens and mosses present in different communities in Ny-Ålesund, Spitsbergen, Arctic

A.	Moss taxa	Substratum	Growth form
1.	*Acarospora heppi* (Nägeli) Nägeli	Rock	Crustose
2.	*Bellemerea cinerorufescens* (Ach.) Clauzade and Roux	Rock	Crustose
3	*Buellia sp.*	Rock	Crustose
4.	*Caloplaca citrina* (Hoffm.) Th.Fr.	Rock, moss	Crustose
5.	*Cetraria aculeata* (Schreb.) Fr.	Moss, soil	Fruticose
6.	*Cetraria fastigiata* (Del. ex Nyl.) Kärnef.	Moss, soil	Fruticose
7.	*Cetraria delisei* (Bory ex Schaer.) Nyl.	Moss, soil	Fruticose
8.	*Cetraria ericetorum* Opiz	Moss, soil	Fruticose
9.	*Cladonia pocillum* (Ach.) Grognet	Moss, soil	Squamulose
10.	*Cladonia pyxidata* (L.) Hoffm.	Moss, soil	Squamulose
11.	*Collema* sp.	Mossy soil	Foliose
12.	*Flavocetraria cucullata* (Bellardi) Karnefelt and Thell	Moss, soil	Fruticose
13.	*Flavocetraria nivalis* (L.) Karnefelt and Thell	Moss, soil	Fruticose
14.	*Lecanora polytropa* (Ehrh.ex Hoffm.) Rabenh.	Rock	Crustose
15.	*Lecanora epibryon* (Ach.) Ach.	Moss, soil	Crustose
16.	*Lecidea lapicida* (Ach.) Ach.	Rock	Crustose
17.	*Ochrolechia frigida* (Sw.) Lynge	Moss, soil	Crustose
18.	*Phaeophysica sciastra* (Ach.) Moberg	Rock	Foliose
19	*Physconia muscigena* (Ach.) Poelt	Soil, moss	Foliose
20.	*Parmelia saxatilis* (L.) Ach.	Rock, moss	Foliose
21.	*Pseudephebe pubescens* (L.) M. Choisy	Rock, moss	Fruticose
22.	*Physcia caesia* (Hoffm.) Fürnr.	Rock, moss	Foliose
23.	*Physcia dubia* (Hoffm.) Lettau	Rock, moss	Foliose
24.	*Rhizocarpon geographicum* (L.) DC.	Rock	Crustose
25.	*Stereocaulon alpinum* Laurer	Moss, soil	Fruticose (Pseudopodetia)
26.	*Stereocaulon vesuvianum* Pers.	Moss, soil	Fruticose (Pseudopodetia)
27.	*Umbilicaria kraschenninikovii* (Savicz) Zahlbr.	Rock	Foliose
28.	*Umbilicaria cylindrica* (L.) Delise ex Duby	Rock	Foliose
29.	*Umbilicaria polyphylla* (L.) Baumg.	Rock	Foliose
30.	*Umbilicaria virginis* Schaer.	Rock	Foliose
31.	*Umbilicaria hyperborea* (Ach.) Hoffm.	Rock	Foliose
32.	*Umbilicaria proboscidea* (L.) Schrad.	Rock	Foliose
33.	*Xanthoria elegans* (Link) Th.Fr.	Rock	Foliose
34.	*Xanthoparmelia sp.*	Rock, soil	Foliose
B.	**Moss taxa**	**Substratum**	**Growth form**
1.	*Aulacomnium palustre* (Hedw.) Schwäegr.	Moist gravel, sand	Axis with leaflet
2.	*Bryum arcticum* Bruch & Schimp	Sandy soil	Axis with leaflet

(continued)

Table 7.1 (continued)

A.	Moss taxa	Substratum	Growth form
3.	*Campylium polygamum* (Schimp.) J. Lange and C. Jensen	Silty soil	Axis with leaflet
4.	*Hymenoloma crispula* (Hedw.) Ochyra	Moist sand	Axis with leaflet
5.	*Kiaeria starkei* (F. Weber and D. Mohr) I. Hagen	Moist sand	Axis with leaflet
6.	*Meesia hexasticha* Funck	Sandy soil	Axis with leaflet
7.	*Niphotrichum panschii* (Müll.Hal.) Bednarek-Ochyra and Ochyra	Sandy soil	Axis with leaflet
8.	*Oncophorus wahlenbergii* Brid.	Calcareous soil	Axis with leaflet
9.	*Sanionia georgicouncinata* (Müll.Hal.) Ochyra and Hedenäs	Sandy-Silty soil	Axis with leaflet
10.	*Sanionia uncinata* (Hedw.) Loeske	Sandy-silty soil	Axis with leaflet
11.	*Tetraplodon mnioides* (Hedw.) Bruch and Schimp	Sand	Axis with leaflet
12.	*Tortella fragilis* (Hook. & Wils.) Limpr.	Sand	Axis with leaflet

different communities in Ny-Ålesund: (a) barren areas, (b) open plains, (c) bird habitats, and (d) wetlands.

7.2.1 Barren Areas

The communities occurring in barren areas are very poor. It mainly consists of epilithic crustaceous lichens and scarce mosses. The barren areas could be delineated as mountains, gracious snow beds, and marine terraces under the influence of strong winds. Soils of these areas are dry as they are unable to hold water. The communities studied in the barren areas comprise of *Acarospora heppi, Bellemerea cinereorufercens, Buellia* sp., *Lecanora polytropa, L. polytropa, Lecidea lapicida, Ochrolechia frigida, Phaeophysica sciastra, Pseudephebe pubescens, Rhizocarpon geographicum, Umbilicaria kraschenninikovii, U. cylindrica* and *U. virginis.* Flowering plants *Salix* sp. and *Saxifraga* sp. along with moss species *Aulacomnium palustre, Niphotrichum panschi*, are growing in association with lichens in the barren areas.

7.2.2 Open Plains

The communities occurring on open plains are one of the most beautiful communities of the tundra, with an appearance of a patterned carpet. Communities found growing abundantly in this habitat comprise of *Cetraria aculeata, C. fastigiata, C. delisei, C. ericetorum, Cladonia pocillum, C. pyxidata, Collema*

Fig. 7.2 (**a**) Arctic tundra (Ny-Ålesund), (**b**) lichen-tundra, (**c**) moss-tundra, (**d**) moss tundra is being invaded by macrofungi and angiosperm, (**e**) lichen and moss communities near glacier showing the invasion of a flowering plant, (**f**) flowering plant community with a few mosses

sp., Flavocetraria cucullata, F. nivalis, Lecanora epibryon, Ochrolechia frigida, Physconia muscigena, Stereocaulon alpinum and *S. vesuvianum*. Most of the lichens are substrate specific while some lichens were found growing on any available substrate. The composition of species varies according to the moistness or dryness of the localities. Four species of mosses: *Bryum arcticum, Kiaeria starkei, Hymenoloma crispula* and *Sanionia uncinata* are common in plain areas. Purple flowering plants *Saxifraga* and grasses grew in association with lichens. The organic matter comprising of dead plants, humus is frequent in the coastal plains, coastal lakes, supporting the colonization and succession in the area.

7.2.3 Bird Habitats

The communities occurring at bird habitats are nitrophilous growing on the foot of slope, boulders, and stones. Bird's droppings provide abundant access to mineral nutrients, particularly nitrogen and phosphorus. The communities studied comprise of *Buellia sp., Caloplaca citrina, Lecidea lapicida, Physcia caesia, P. dubia, Pseudephebe pubescens, Umbilicaria proboscidea and Xanthoria elegans*. There is a mixture of mosses, herbs and grasses with the lichens at the foot of slopes because bird excrement makes them a more favorable place for their growth. Seeds from some of the flowering plants provide food for migrating birds.

Lichens dominate the tundra as major primary producers. Associations among lichens are a common observation in Ny-Ålesund. About two to six different lichen species were found growing on a piece of substratum collected for the study. It clearly indicates the luxuriant growth of lichens on the island. Though, identical requirement for nutrients and microclimatic conditions may result in such an association, an in-depth study is required to draw any further conclusions. *Cetraria fastigiata* with *Flavocetraria cucullata, Lecanora polytropa* with *Lecidea lapicida, Umbilicaria proboscidea* with *Pseudephebe pubescens, Bellemerea cinerorufescens* with *Umbilicaria kraschenninikovii, Physcia caesia* with *Xanthoria elegans, Physconia muscigena* with *Caloplaca citrina, Cladonia pyxidata* with *Ochrolechia frigida* are the most common associations seen in the study area. The communities of these lichens in different habitats appear as lichen-tundra (Fig. 7.2b).

7.2.4 Wetlands

The communities occurring at wetland are dominated by mosses. The mosses grow on moist rock, gravel, sand and silt. The communities studied here cover maximum area and are found to comprise of *Aulacomnium palustre, Bryum arcticum, Campylium polygamum, Hymenoloma crispula, Kiaeria starkei, Meesia hexasticha, Oncophorus wahlenbergii, Sanionia georgicouncinata* and *Tortella fragilis*. Among these, *Aulacomnium palustre* and *Oncophorus wahlenbergii* are widely distributed while *Kiaeria starkei* and *Meesia hexasticha* are very rare in the area. The community of these species appears as moss-tundra in the area. Tundra-moss grows in areas where there is some soil and sufficient water. They form a thick soft and squishy cushion like greenish terrain in the area (Fig. 7.2c). However, at many places various types of fungi, grasses and herbs have begun to invade the mosses during summer period (Fig. 7.2d).

The spatial pattern of communities shows that their development depends on climatic factors such as wind, snow cover, soil condition, water, geological substratum, and occurrence of colonies of birds. Some communities occupy great areas while others are very small and some are mosaic.

7.3 Impact of Climate Change

Climate change is beginning to cause large-scale changes in Ny-Ålesund. Due to increase in summer temperature and glacial retreat in particular, fresh area is exposed for colonisation and succession of new community in the areas (Fig. 7.2e). Presently, the tundra is home to several species of lichens and mosses but the warming is causing the lichen and moss to disappear due to competition and dominance by angiosperms (Fig. 7.2f). An increased temperature leads to a rise in the biomass of flowering plants; and a subsequent drop in the biomass of lichens and mosses (Press et al. 1998). In general, the vascular plants are unable to go through the whole life cycle in severe cold climate. It is predicted that rising temperatures will kill off lichens and mosses up to 40% of the tundra, allowing angiosperm plant species to grow in the warmer climate. The Arctic plants flower during the months of July and August when temperature reaches +2.5 to +5 °C (Dubiel 1988). There is now ample evidence of the ecological impacts of recent climate change in Ny Alesund Arctic. Temperature of Ny-Ålesund is rising and causing the invasion of flowering plants into lichen and moss communities. This process is threatening the existing lichen and moss population and further, it will lead to the dominance of flowering plants in tundra. Changes in community structure can profoundly affect the pattern of the ecosystem's carbon cycle, especially when the shift occurs from one species to another with different physiological properties (Cornelissen et al. 2001).

7.4 Concluding Remarks

Although we are at an early stage in the projected trends of global warming, ecological responses to recent climate change are visible from species to the community levels. An in-depth study encompassing biogeochemical parameters is required to corroborate the observations and draw further conclusions on climate change and resultant ecological succession.

Acknowledgement We are thankful to the Secretary, Ministry of Earth Sciences, Government of India for support.

References

Cornelissen JHC, Callaghan TV, Alatalo JM, Michelsen A, Graglia E, Hartley AE, Hik DS, Hobbie SE, Press MC, Robinson CH, Henry GHR, Shaver GR, Phoenix GK, Gwynn JD, Jonasson S, Chapin FS III, Molau U, Neill C, Lee JA, Melillo JM, Sveinbjornsson B, Aerts R (2001) Global change and Arctic ecosystems: is lichen decline a function of increases in vascular plant biomass? J Ecol 89:984–994

Dillenius JJ (1741) Historia muscorum. Oxonii XVI + 576 pp., 85 Pl
Dubiel E (1988) Ecological observations on vascular plants in the NW Sörkapp land (Spitsbergen). Uniw Jagiell Prace Bot Zeszyt 22:39–46
Elvebakk A (1984) Contributions to the lichen flora and ecology of Svalbard, arctic Norway. Bryologist 87:308–313
Elvebakk A, Hertel H (1996) Lichens. In: Elvebakk A, Prestrud P (eds) A catalogue of Svalbard plants, fungi, algae, and cyanobacteria. Norsk Polarinstitute, Oslo, pp 271–348
Frisvoll AA, Elvebakk A (1996) Part 2. Bryophytes. In: Elvebakk A, Prestrud P (eds) A catalogue of Svalbard plants, fungi, algae and cyanobacteria. Norsk Polarinstitute, Oslo, pp 57–172
Hafellner J (1982) Flechtenfunde am Bockfjord, Spitzbergen. Ergebnisse der osterreichischen Spitzbergen-Expedition 1979, I. Phyton 22:23–50
Kuc M (1973) A review of the mosses of the Svalbard. Rev Bryol Lichenol 39:401–472
Lindberg SO (1883) Kritisk granskning af mossorna uti Dillenii Historia muscorum. J. C. Frenckell. & Son, Helsingfors, 59 p
Lynge B (1938) Lichens from the west and north coasts of Spitsbergen and the North –East Land collected by numerous expeditions. I. The macrolichens. Skr. Norske Vidensk.-akad. Oslo. I. Mat. –Nat. Vitensk. Kl. 6, 1–136, Pl
Martens F (1675) Spitzbergische oder Groenlandische Reise Beschreibung gethan im Jahr 1671. Dritter Theil. Von den Pflanzen so ich in Spitsbergen gefunden. pp 41–51. Hamburg. Translated in Martens F (1855) Voyage to Spitzbergen. Part the third. Of the plants of Spitzbergen. In: White A (ed) A collection of documents on Spitzbergen and Greenland, comprising a translation from F. Martens' voyage to Spitzbergen. Work issued by the Haklyut Society 18. London, pp 45–65
Mattick F (1949) Die Flechten Spitzbergens. Polarforschung 19:261–273
Ochyra R, Singh SM, Bednarek-Ochyra H (2009) Meesia hexasticha (Funck) Bruch. In: Blockeel TL (ed) New national and regional bryophyte records, 20. J Bryol 31:56
Olech M (1987) Materials to the lichen flora of Hornsund (SW Spitsbergen). Zesz Nauk Uniw Jagiell Prace Bot 19:165–168
Olech M, Alstrup V (1989) Lichens new to Spitsbergen. Graphis Scripta 2:146–148
Phips CI (1777) Reise nach dem Nordpol. Auf Befehl Ihro Konigl. Grossbrittannischen Majestat. Unternommen im Iahr 1773. Bern
Press MC, Potter JA, Burke MJW, Callaghan TV, Lee JA (1998) Responses of a Subarctic dwarf shrub heath community to stimulated environmental change. J Ecol 86:315–327
Søchting U, Olech M (1995) The lichen genus Caloplaca in polar regions. Lichenologist 27:463–471
Thomson JW (1984) American Arctic lichens. I. The macrolichens. Columbia University Press, New York
Thomson JW (1997) American Arctic lichens. 2. The microlichens. University of Wisconsin Press, Madison

Chapter 8
High Resolution Southwest Monsoon Reconstruction for the Past ~2,800 Years: Wind Versus Precipitation

Manish Tiwari

8.1 Introduction

Most of the earlier studies on paleomonsoon variations are concentrated on the western Arabian Sea, as it undergoes intense biogeochemical changes during the Southwest (SW) monsoon (Nair et al. 1989) that are easily detectable. But as that area is dominated by wind induced upwelling, hence these studies have focused on its effects on various processes such as influx of nutrients and colder water from depth resulting in drastic changes in productivity, abundance of particular species of microorganisms, aeolian dust content, isotopic content of surface water etc. (e.g. Clemens et al. 1991; Anderson and Prell 1993; Sirocko et al. 1993; Naidu et al. 1993; Reichart et al. 2002; Gupta et al. 2003, 2005; Saher et al. 2007; Tiwari et al. 2010 and references therein) and not on the precipitation signal. Stronger SW monsoon winds do not necessarily mean higher monsoon precipitation as, the latter depends on various other parameters such as the moisture content of the air-masses, which is probably influenced by the sea surface temperature (SST) of southern Indian Ocean (Clemens et al. 1996) and air parcel convergence and convection (Gadgil 2003). Therefore, it becomes very important to test how the wind speed correlates with the precipitation and how this relationship varied in the past. The eastern Arabian Sea offers such an opportunity as it receives abundant fresh water as either direct overhead precipitation or surface runoff from the adjacent hills, present along the western Indian coast (Western Ghats), which induce intense orographic precipitation during the SW monsoon (June–September). The influx of copious amounts of fresh water into the coastal eastern Arabian Sea reduces the sea surface salinity (SSS) that is reflected in the various proxies that ultimately get preserved in the sea sediments.

M. Tiwari (✉)
National Centre for Antarctic and Ocean Research, Vasco-da-Gama 403804, India
e-mail: manish@ncaor.org

R. Sinha and R. Ravindra (eds.), *Earth System Processes and Disaster Management*, Society of Earth Scientists Series 1, DOI 10.1007/978-3-642-28845-6_8,

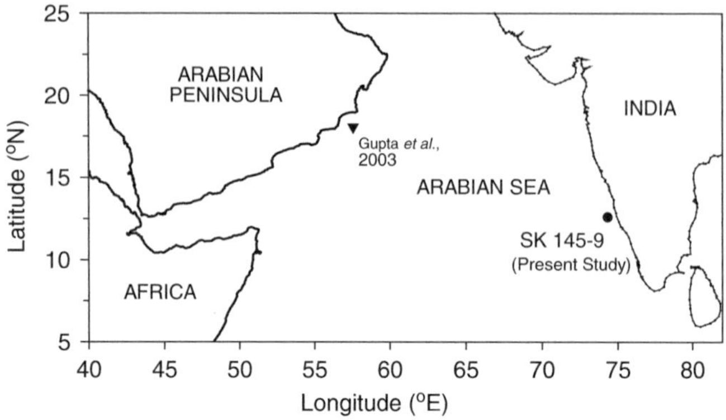

Fig. 8.1 Location of the core SK 145-9 of the present study in the eastern Arabian Sea that records the SW monsoon precipitation signal. The core location of the Gupta et al. (2003) from western Arabian Sea, used for comparison purpose in this study, is also shown that records SW monsoon wind variability

In comparison to western Arabian Sea, only a few studied this aspect in the eastern Arabian Sea (Sarkar et al. 2000; Thamban et al. 2001; Agnihotri et al. 2003; Tiwari et al. 2006; Chauhan et al. 2009; Govil and Naidu 2010). For the present study, a sediment core, namely SK 145-9, from the eastern Arabian Sea is used. It is obtained from a shallow depth experiencing faster sedimentation that would provide a high time resolution comparable to studies from the western Arabian Sea. This would help in the comparative study of past changes on two different aspects of the monsoon: (1) wind induced upwelling and productivity (2) rainfall and runoff to the ocean and its effect on surface salinity. Furthermore, because of the high sedimentation rate and the subsequent high resolution sampling, it would aid in delineating centennial and sub-centennial scale variations in the monsoon rainfall.

8.2 Materials and Methods

A 252 cm long sediment core (SK 145-9) was collected from the eastern Arabian Sea (location: 12.6°N, 74.3°E) from a water depth of 400 m off the Mangalore coast (location shown in Fig. 8.1).

The top 50 cm have been sampled at every cm after which the sampling was done at every two cm. The top 50 cm of this core is considered for further studies as it has been sampled at high-resolution and therefore will aid in understanding sub-centennial scale variability. In the studied length of 50 cm, four AMS radiocarbon dates were obtained on planktic foraminiferal separates (comprising *Globigerinoides ruber, Globigerinoides sacculifer, Orbulina universa* and *Neogloboquadrina dutertrei*), spanning ~2,860 cal years Before Present (BP) with an average resolution of ~57 years per cm (see Table 8.1, Fig. 8.2). The intermediate depths have been assigned dates based on sedimentation rates as shown in Fig. 8.2. The dating has been

Table 8.1 Calibrated radiocarbon ages in the top 50 cm of the core SK 145-9

Laboratory no.	Sample depth (cm below sea floor)	AMS sample no.	Radiocarbon age (^{14}C yr BP)	Calibrated age (cal yr BP)
T11520B	0–2	AA36397	844 ± 36	410 ± 80
T11521A	23–25	AA36398	1,899 ± 56	1,330 ± 80
T14807A	40–41	AA43871	2,506 ± 39	2,030 ± 70
T11523	50–52	AA36400	3,210 ± 41	2,860 ± 70

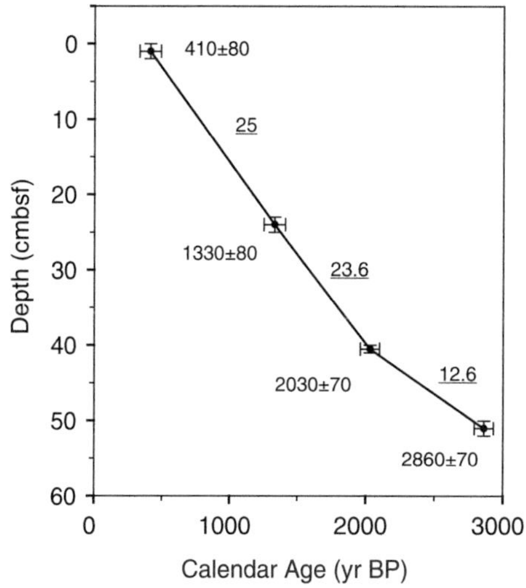

Fig. 8.2 The 'Age-Depth Model' for the core SK 145 9. Four AMS dates with errors are shown along with sedimentation rates (*underlined*) in cm per thousand years; 'cmbsf' denotes 'centimetres below sea floor' while 'yr BP' denotes 'years Before Present'

carried out at the NSF accelerator mass spectrometer (AMS) facility, University of Arizona, USA. The top of the core (viz. 0–2 cm) has a calibrated age of 410 ± 80 cal yr BP. The radiocarbon dates in this core have been converted to calendar ages using the calibration program CALIB 4.1 (INTCAL 98) (Stuiver et al. 1998) with a reservoir age correction of 540 ± 20 years (Southon et al. 2002).

For stable oxygen (δ^{18}O) and carbon (δ^{13}C) isotope analysis, ~5–10 g of sediment samples were dissolved in ~200 mL of distilled water to which was added ~½ cc of 30% H_2O_2 along with ½ spoonful of Calgon powder (sodium hexametaphosphate). It was then warmed at 60–80°C for an hour. Subsequently wet sieving was carried out using sieves of the sizes 250 and 500 μm and the shells in the 250–500 μm size range were collected. The seived shells were then ultrasonically cleaned; oven dried at 90°C and kept in a furnace at 400°C for ~8 h to remove any volatile organic matter still adhering to the shells. δ^{18}O and δ^{13}C analysis has been carried out on 30–35 handpicked shells of the three species of planktic foraminifera viz., *Globigerinoides ruber*, *Globigerinoides sacculifer*, and *Globorotalia menardii* using GEO 20-20 mass spectrometer at Physical Research Laboratory, Ahmedabad. The size range chosen for separating bulk foraminifera was 250–500 μm but while picking foraminifera for isotopic analysis, care was taken to pick the shells falling approximately in the range

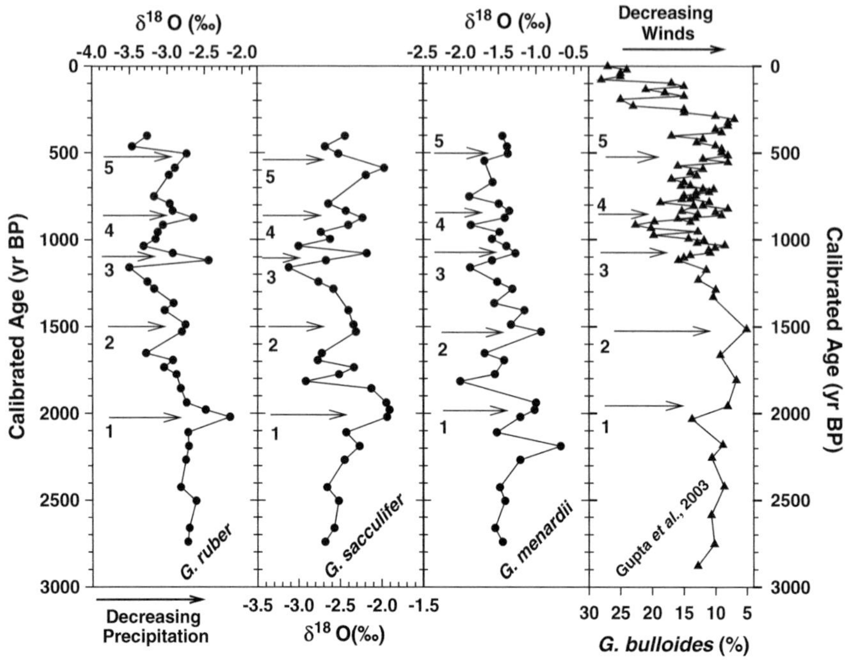

Fig. 8.3 Downcore variations of $\delta^{18}O$ of three species of foraminifera depicting SW monsoon precipitation variability in the core SK 145-9; *numbered arrows* depict arid episodes. The *last panel* shows the *G. bulloides* abundance data from Gupta et al. (2003) that denotes SW monsoon wind variability (modified after Tiwari et al. 2006)

350–450 μm. The external precision on $\delta^{18}O$ and $\delta^{13}C$ measurement is ±0.1‰ and ±0.04‰ respectively (1σ standard deviation) obtained by repeatedly running Z-Carrara as the internal/Lab Standard (no. of repetition = 62). All isotopic values are reported with respect to V-PDB. The reference standard used for normalizing to V-PDB scale is NBS-19 that has $\delta^{18}O$ and $\delta^{13}C$ values of −2.20‰ and +1.95‰ respectively with respect to V-PDB.

8.3 Results

8.3.1 Precipitation Signals as Manifested in Oxygen Isotopes

G. ruber and *G. sacculifer* are surface dwelling species predominantly inhabiting top 25 m and 50 m respectively whereas *G. menardii* is a deeper dwelling species predominantly inhabiting 100–150 m (Bé 1977; Fairbanks et al. 1980). Thus an oxygen isotope variability arising due to any surface processes (e.g. salinity change) will be most pronounced in the surface dwelling species viz. *G. ruber* and *G. sacculifer* and will be subdued in the deeper dwelling species i.e. *G. menardii*.

The first three panels of Fig. 8.3 shows the $\delta^{18}O$ variation in three different species of foraminifera viz. *G. ruber*, *G. sacculifer*, *G. menardii* – all the three species show similar signals that are somewhat restrained in the deeper dwelling one. The oxygen isotope data for *G. ruber* and *G. sacculifer* is taken from Tiwari et al. (2006) for comparison purpose while all other data presented here is unpublished.

The studied species are known to grow predominantly during the SW monsoon months and hence are likely to record the signals arising mainly due to SW monsoon fluctuation (Guptha et al. 1997). The factors controlling the oxygen isotopes in foraminiferal shells are the sea surface salinity (SSS) and sea surface temperature (SST; Shackleton 1967). For the past ~3 ka there has been no salinity fluctuations due to the global ice-volume effect as there were no significant global ice-melting episodes affecting sea level (Fairbanks 1989). Moreover the SST variations in the tropics during the Holocene are very small (~0.5°C, Rostek et al. 1993). In the eastern Arabian Sea, SSS variation is mainly controlled by the variation in the supply of fresh water as surface runoff from the adjacent Western Ghats during the southwest monsoon. Furthermore, a weak upwelling system occurs in the eastern Arabian Sea along the western Indian coast prior to the SW summer monsoon that gets established from February onwards (Shetye 1984). This upwelling is controlled by factors other than the SW monsoon winds. Studies by Shankar and Shetye (1997) and McCreary et al. (1993) have shown that the early upwelling is a result of the remote forcing by winds in the Bay of Bengal and southwest coast of India that generate the northward propagating Kelvin and westward propagating Rossby waves. It is further confirmed by the study carried out by Thamban et al. (2001) on a sediment core near the Cochin coast (southwest continental margin of India). But, as the upwelling is affected by basin-wide remote processes in Arabian Sea and Bay of Bengal (McCreary et al. 1993), the variation in the local upwelling intensity (solely due to SW monsoon intensity) will only affect the interannual SST changes in a small way (Thamban et al. 2001). It can be therefore be assumed safely that the dominant factor controlling the $\delta^{18}O$ signals in the eastern Arabian Sea is the SSS changes induced by the variation in the SW monsoon precipitation. A reduction in SSS occurs due to the influx of large amount of fresh water, depleted in ^{18}O, as surface runoff into the coastal eastern Arabian Sea during intense SW monsoon precipitation events. In the eastern Arabian Sea, for every per mil decline in salinity, the $\delta^{18}O$ value decreases by 0.33‰ (Duplessy et al. 1981; Sarkar et al. 2000). Thus a depleted $\delta^{18}O$ signal indicates enhanced southwest monsoon precipitation whereas an enriched $\delta^{18}O$ signal points towards reduced precipitation due to weaker southwest monsoon (indicated by arrows in Fig. 8.3). Five such arid periods have been identified in the oxygen isotope record, which are centered at ~2,000, 1,500, 1,100, 850 and 500 cal yr BP (arrows 1–5 in Fig. 8.3). The first arid event is observed at ~2,000 cal yr BP with all the three species exhibiting high $\delta^{18}O$ values indicating a major period of aridity during that time. Another high $\delta^{18}O$ value is observed in *G. menardii* record at ~2,200 cal yr BP. As this event is not registered in the predominantly surface dwelling species and is found only in predominantly thermocline dwelling species that may get affected by deeper processes (ocean undercurrents), it is not considered as a low precipitation event.

Another arid event of a smaller magnitude is observed at ~1,500 cal yr BP (shown by arrow-2), shown by all the three species of foraminifera. Thereafter, precipitation intensity strengthens up to ~1,200 cal yr BP after which it shows a sudden decline centered at ~1,100 cal yr BP (shown by arrow-3) and then monsoon intensified rapidly. Other two episodes of precipitation reduction are observed at the ~850 cal yr BP and ~500 cal yr BP (arrows 4 and 5). All these events of aridity/ reduced precipitation are reflected in all the three species of the foraminifera, confirming their genuineness.

8.3.2 Productivity as Manifested in Carbon Isotopes

The $\delta^{13}C$ values of foraminifera show the carbon isotopic composition of seawater, which in turn is predominantly governed by the organic productivity on short timescales. Kinetic isotope effects during photosynthesis cause preferential uptake of lighter isotope (^{12}C) in the organic matter making the ambient Total Dissolved Inorganic Carbon (DIC) of seawater enriched in heavier isotope (higher $\delta^{13}C$ value). Foraminifera incorporate this signature of carbon isotopes when they precipitate their shells in equilibrium with seawater following the reaction: Ca^{2+} (aq.) + $2HCO_3^-$ (aq.) $\leftrightarrow$ $CaCO_3$ (s) + H_2O + CO_2 (g). Thus $\delta^{13}C$ values of foraminifera reflect productivity. Another competing influence on carbon isotopic composition of foraminifera is that of upwelling which brings ^{13}C depleted water to the surface and thus reduces the $\delta^{13}C$ of the foraminifera (Kroopnick 1985). But as discussed earlier, the effect of upwelling along the west coast of India is very less and is governed by processes other than SW monsoon intensity; hence it will have negligible effect when compared the predominant process of productivity. Figure 8.4 shows the temporal variations in $\delta^{13}C$ in three species of foraminifera viz. *G. ruber*, *G. sacculifer* and *G. menardii*.

The carbon isotope record shows several fluctuations but there are prominent peaks that are identifiable in all the three species. Such peaks have been numbered 1–5 that matches considerably well with the arid events as deciphered from the $\delta^{18}O$ record. The arrows points towards decreasing productivity and it's evident that arid episodes are accompanied by marked decline in productivity. Productivity is intrinsically coupled with monsoon; SW monsoon winds might cause mixed layer deepening, which injects nutrient rich waters into the surface layer enhancing surface productivity. Another possibility is increased productivity due to nutrients derived from land in the form of surface runoff. Conversely lower productivity may result during periods of weak SW monsoon. In fact, present day data from the Joint Global Ocean Flux Studies (JGOFS-India) program during the 1994–1995 shows that the primary productivity reaches ~0.6 $gC/m^2/d$ (gram carbon/meter^2/day) during the SW monsoon whereas it is only ~0.3 and ~0.2 $gC/m^2/d$ during the NE

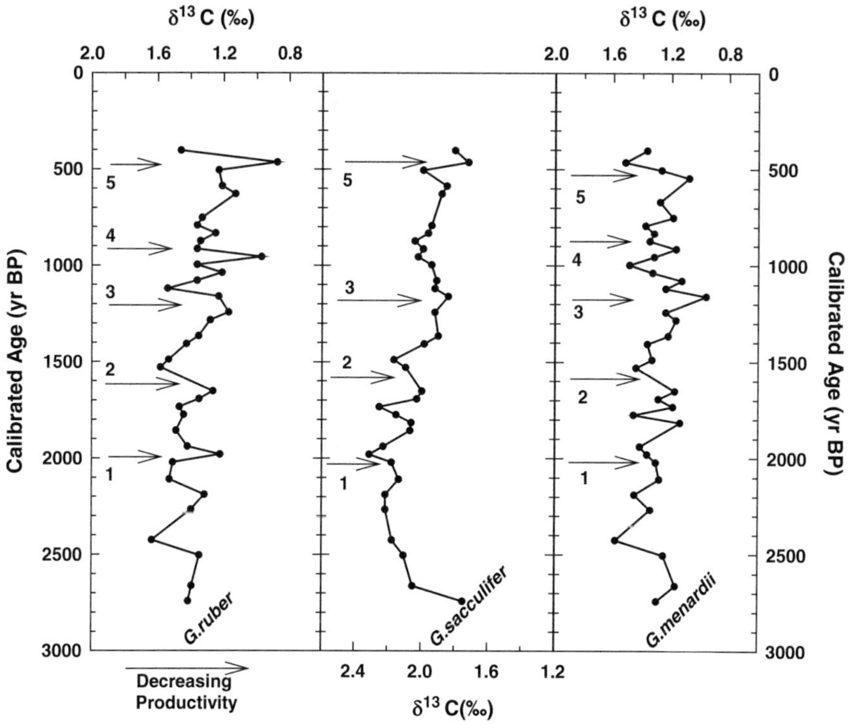

Fig. 8.4 Downcore variations of $\delta^{13}C$ of three species of foraminifera depicting productivity variability in the core SK 145-9, *numbered arrows* depict arid episodes as identified in the oxygen isotope record

monsoon and intermonsoon, respectively (Bhattathiri et al. 1996). It implies clearly that SW monsoon affects productivity in the eastern Arabian Sea and hence productivity can act as an excellent proxy for past SW monsoon variability in this region.

The productivity shows decline during ~2,000, 1,600, 1,200, 900, 500 cal yr BP (arrows 1–5) that correlates well, within the radiocarbon dating errors, with the arid events identified in oxygen isotope record. Again it is important to stress the point that the signals are valid and strong as they are observed in three different species of foraminifera living in different water depths. Although a few exception exists such as arid/low-productivity event at arrow-4 is missing in the *G. sacculifer* record. But arrow-4 event is considered as real because it is accompanied with an arid episode as deciphered from an entirely different oxygen isotope proxy.

These concurrent evidences from high-resolution oxygen and carbon isotope records of three different species of foraminifera lead us to identify five episodes of strong decline in SW monsoon precipitation at ~2,000, 1,500, 1,100, 850 and 500 cal yr BP.

8.4 Discussion

The low-*precipitation* events as identified above offers us an opportunity to compare this record with other high-resolution records western/northern Arabian Sea that may yield SW monsoon *wind* intensity record. This would answer the long-standing question raised by paleoclimatologists studying land based records that whether the SW monsoon wind records from western Arabian Sea (near Arabian peninsula) truly represent precipitation over Indian landmass. One such high-resolution record depicting wind strength is available from ODP Site 723A (Gupta et al. 2003, 2005), which is based on abundance of *Globigerina bulloides* (shown in the extreme right panel in Fig. 8.3). *G. bulloides* is a temperate water species and can occur in tropics where cooler water is present due to upwelling. Thus their abundance as expressed in percentage of total planktic foraminifera is an excellent indicator of upwelling, which in turn is controlled by the wind strength (Anderson et al. 2002). The 2,000 cal yr BP arid event observed in the precipitation record is also observed in the % *G. bulloides* (Gupta et al. 2003) record that shows a decrease at that time indicating reduced wind intensity. Zonneveld et al. (1997) obtained a core from the Somalian coast, western Arabian Sea and found a decline in the upwelling intensity ~2,000 cal yr BP on the basis of dinoflagellates cysts and inferred that SW monsoon wind strength declined during that time. In our data, another arid event of a smaller magnitude is observed at 1,500 cal yr BP (depicted by arrow-2), shown by all the three species of foraminifera. At that time % *G. bulloides* exhibits minimum abundance (~5%) in the western Arabian Sea (Gupta et al. 2003) pointing towards an extreme reduction in upwelling and hence the wind strength. But this minimum wind strength does not correspond to minimum precipitation, which happens at ~2,000 cal yr BP as evident by the maximum $\delta^{18}O$ values during that time. Thereafter, precipitation intensity strengthens up to ~1,200 cal yr BP after which it shows a sudden decline centered at ~1,100 cal yr BP (shown by arrow-3 in Fig. 8.3) and then monsoon intensified as rapidly. This sudden reduction and subsequent enhancement in monsoon intensity is also observed in the wind intensity record of the western Arabian Sea as shown by sharp decline in % *G. bulloides* during the above period. Other episodes of precipitation reduction, observed at the ~850 cal yr BP and 500 cal yr BP (arrows 4 and 5), are also accompanied by a corresponding reduction in the SW monsoon wind intensity (Gupta et al. 2003) in western Arabian Sea (Fig. 8.3).

Comparison with other records of precipitation from diverse regions of Arabian Sea and surrounding landmass yield good correspondence with the aridity events observed in the present study. A reduction in monsoon is also indicated by a hiatus in the growth of speleothem in Oman from 2,500 to 1,500 cal yr BP (Fleitmann et al. 2003). Staubwasser et al. (2003) found reduced precipitation at ~2,000 cal yr BP on the basis of increasing $\delta^{18}O$ in *G. ruber* from a core off the Pakistan coast, which records signal of Indus discharge governed by SW monsoon precipitation in its catchment area. Similarly, Sarkar et al. (2000) and Thamban et al. (2001) have taken sediment cores from the eastern Arabian Sea near to the Mangalore coast and

Cochin coast respectively and measured $\delta^{18}O$ values in planktic foraminifera. Sarkar et al. (2000) measured $\delta^{18}O$ and $\delta^{13}C$ of the two species of planktic foraminifera namely *G. sacculifer* and *G. menardii* in their core 3268G5. The time period they studied was the last 10,000 years and inferred the excess of evaporation over precipitation (E–P). Likewise, Thamban et al. (2001) determined the $\delta^{18}O$ and $\delta^{13}C$ in planktic and benthic foraminifera in the core GC-5. They reported significant fluctuation in $\delta^{18}O$ during Holocene due to changes in sea surface characteristics relatable to variations in SW monsoon over land. Although their cores have coarser resolution, still they exhibit reduction in SW monsoon precipitation during ~2,000 yr BP. Recently Chauhan et al. (2009) obtained a core from the southeastern Arabian Sea and measured oxygen isotopes and other geochemical proxies. They also report a major arid event centered at ~2,000 cal yr BP. Thus it is clear that the 2,000 cal yr BP arid event was a widespread episode of aridity, which was recorded in diverse proxies in the Arabian Sea.

Thereafter, other four low precipitation events centered at ~1,500, 1,100, 850, 500 cal yr BP are observed in varied records such as speleothems (Yadava and Ramesh 2005), varves (von Rad et al. 1999; Luckge et al. 2001), oxygen isotopes (Staubwasser et al. 2003; Chauhan et al. 2009) and the Oman stalagmites (Fleitmann et al. 2003). Chauhan et al. (2009) obtained a core from near the southern tip of India and reconstructed SW monsoon precipitation variability at sub-centennial resolution. They identified three prominent arid periods at ~2,000, ~1,100, ~500 cal yr BP that matches very well with the results of the present study. Yadava and Ramesh (2005) studied the speleothems from several regions of India such as Chhattisgarh, Chitrakoot district of UP, Uttar Kannada district of Karnataka, Pithoragarh district of western Himalaya and reconstructed the SW monsoon precipitation over land at very high resolution (sub-decadal to decadal scale). Although it becomes difficult to compare with their record with much higher resolution but they also observed rapid increase in $\delta^{18}O$ values at ~2,000, 1,500 ^{14}C yr BP and 1,100 cal yr BP indicating major period of aridity as found in the present study. von Rad et al. (1999) and Luckge et al. (2001) studied varve thickness along with $\delta^{18}O$ and other geochemical proxies ($CaCO_3$ content, Ti/Al, K/Al) in a core from the northern Arabian Sea at very high resolution (sub-decadal) spanning a period of 5,000 years. During the past 3,000 years, the time period of interest vis-à-vis the present study, they found period of aridity at ~2,000, 1,750, 1,000, 600 cal yr BP that match well with the present study with in the radiocarbon dating limits. Fleitmann et al. (2003) in their stalagmite record from Oman, after the hiatus from 2,000 to 1,500 cal year BP, also observed positive incursions in $\delta^{18}O$ values (depicting aridity) at ~1,400, 1,000, 500 cal yr BP that agrees well with the present record.

Thus, the above discussion proves that the periods of aridity observed in the eastern Arabian Sea were widespread and were reflected in diverse proxies from different regions around it. Also, wind speed indicators from the western Arabian Sea exhibit a good correlation with the precipitation signals from the eastern Arabian Sea at least during the last 2,800 years as evident from the simultaneous reduction observed in both the records. Present day climatological studies have shown that SW monsoon winds initiates due to the land sea pressure contrast but are sustained through the strengthening of the low pressure cell over land mass through the latent heat released during condensation/precipitation. Thus precipitation intensity modulates the wind strength. Our study provides a paleoclimatic evidence for the same; whenever precipitation reduced over Indian landmass, the SW monsoon winds weakened in western Arabian Sea.

8.5 Conclusion

A prominent arid period is observed at ~2,000 yr BP. Thereafter several arid periods are observed at ~1,500, ~1,100, ~850 and ~500 yr BP, which are also seen in other proxy records from Arabian Sea. SW monsoon wind intensity exhibits excellent correlation with the SW monsoon precipitation. It also provides the paleoclimatic evidence for the fact that SW monsoon governs the wind intensity *via* latent heat release. Furthermore, this study shows that carbon isotope variability ($\delta^{13}C$) of foraminifera in eastern Arabian Sea can act as an excellent indicator of productivity, which in turn is governed not only by wind induced mixing but also by the nutrient influx *via* surface runoff. Productivity is intrinsically linked with the SW monsoon strength; low productivity as evident from $\delta^{13}C$ record occurs during periods of aridity.

Acknowledgement MT thanks ISRO-GBP for funding and Sri Rasik Ravindra, Director-NCAOR for support; this is NCAOR Contribution no. 015/2011.

References

Agnihotri R, Bhattacharya SK, Sarin MM, Somayajulu BLK (2003) Changes in the surface productivity and subsurface denitrification during the Holocene: a multiproxy study from the eastern Arabian Sea. Holocene 13(5):701–713

Anderson DM, Prell WL (1993) A 300 kyr record of upwelling off Oman during the Late Quaternary: evidence of the Asian Southwest monsoon. Paleoceanography 8:193–208

Anderson DM, Overpeck JT, Gupta AK (2002) Increase in the Asian SW monsoon during the past four centuries. Science 297:596–599

Bé AWH (1977) An ecological, zoogeographic and taxonomic review of recent planktonic foraminifera. In: Ramsey ATS (ed) Oceanic micropaleontology. Academic, London, p 100

Bhattathiri PMA, Pant A, Sawant S, Gauns M, Matondakar SGP, Mohanraju R (1996) Phytoplankton production and chlorophyll distribution in the eastern and central Arabian Sea in 1994–1995. Curr Sci 71(11):857–862

Chauhan OS, Vogelsang E, Basavaiah N, Kader USA (2009) Reconstruction of the variability of the southwest monsoon during the past 3 ka, from the continental margin of the southeastern Arabian Sea. J Quat Sci 25(5):798–807

Clemens SC, Prell W, Murray G, Shimmield G, Weedon G (1991) Forcing mechanisms of the Indian Ocean monsoon. Nature 353:720–725

Clemens SC, Murray DW, Prell W (1996) Non-stationary phase of the Plio-Pleistocene Asian monsoon. Science 274:943–948

Duplessy JC, Bé AWH, Blanc PL (1981) Oxygen and carbon isotopic composition of planktonic foraminifera in the Indian Ocean. Palaeogeogr Palaeoclimatol Palaeoecol 33:9–46
Fairbanks RG (1989) A 17,000-year glacio-eustatic sea level record: influence of glacial melting rates on the Younger Dryas event and deep ocean circulation. Nature 342:637–642
Fairbanks RG, Wiebe PH, Bé AWH (1980) Vertical distribution and isotopic composition of living planktonic foraminifera in the western North Atlantic. Science 207:61–63
Fleitmann D, Burns SJ, Mudelsee M, Neff U, Kramers J, Mangini A, Matter A (2003) Holocene forcing of the Indian Monsoon recorded in a stalagmite from southern Oman. Science 300:1737–1739
Gadgil S (2003) The Indian monsoon and its variability. Annu Rev Earth Planet Sci 31:429–467
Govil P, Naidu PD (2010) Evaporation-precipitation changes in the eastern Arabian Sea for the last 68 ka: implications on monsoon variability. Paleoceanography 25. doi:10.1029/2008PA001687
Gupta AK, Anderson DM, Overpeck JT (2003) Abrupt changes in the Asian southwest monsoon during the Holocene and their links to the North Atlantic Ocean. Nature 421:354–357
Gupta AK, Das M, Anderson DM (2005) Solar influence on the Indian summer monsoon during the Holocene. Geophys Res Lett 32:L17703
Guptha MVS, Curry WB, Ittekkot V, Murlinath AS (1997) Seasonal variation in the flux of planktic foraminifera: sediment trap results from the Bay of Bengal, Northern Indian Ocean. J Foramin Res 27:5–19
Kroopnick PM (1985) The distribution of ^{13}C of $\Sigma CO2$ in the world oceans. Deep-Sea Res 32:75–84
Luckge A, Rolinski AD, Khan AA, Schulz H, von Rad U (2001) Monsoonal variability in the northeastern Arabian Sea during the past 5000 years: geochemical evidence from laminated sediments. Palaeogeogr Palaeoclimatol Palaeoecol 167:273–286
McCreary JP, Kundu PK, Molinari RL (1993) A numerical investigation of dynamics, thermodynamics and mixed layer processes in the Indian Ocean. Prog Oceanogr 31:181–224
Naidu PD, Malmgren BA, Bornmalm L (1993) Quaternary history of the calcium carbonate fluctuations in the western equatorial Indian Ocean (Somali basin). Palaeogeog Palaeoclimatol Palaeoecol 103:21–30
Nair RR, Ittekkot V, Ramaswamy SJ, Haake B, Degens ET, Desai BN, Honjo S (1989) Increased particle flux to the deep ocean related to monsoons. Nature 338:749–751
Reichart GJ, Schenau SJ, de Lange GJ, Zachariasse WJ (2002) Synchroneity of oxygen minimum zone intensity on the Oman and Pakistan margins at sub-Milankovitch time scales. Mar Geol 185:403–415
Rostek F, Ruhland G, Bassinot FC, Muller PJ, Labeyrie LD, Lancelot Y, Bard E (1993) Reconstructing sea surface temperature and salinity using $\delta^{18}O$ and alkenone records. Nature 364:319–321
Saher MH, Jung SJA, Elderfield H, Greaves MJ, Kroon D (2007) Sea surface temperatures of the western Arabian Sea during the last deglaciation. Paleoceanography 22:PA2208
Sarkar A, Ramesh R, Somayajulu BLK, Agnihotri R, Jull AJT, Burr GS (2000) High resolution Holocene monsoon record from the eastern Arabian Sea. Earth Planet Sci Lett 177:209–218
Shackleton NJ (1967) Oxygen isotope analyses and Pleistocene temperatures reassessed. Nature 215:15–17
Shankar D, Shetye SR (1997) On the dynamics of the Lakshadweep high and low in the southeastern Arabian Sea. J Geophys Res 102:12551–12562
Shetye SR (1984) Seasonal variability of the temperature field off the southwest coast of India. Proc Indian Acad Sci 93(4):399–411
Sirocko F, Sarnthein M, Erlenkreuser H, Lange H, Arnold M, Duplessy JC (1993) Century scale events in monsoon climate over the past 24,000 years. Nature 364:322–324
Southon J, Kashgarian M, Fontugne M, Metivier B, Wyss Yim W-S (2002) Marine reservoir corrections for the Indian Ocean and Southeast Asia. Radiocarbon 44:167–180

Staubwasser M, Sirocko F, Grootes PM, Segl M (2003) Climate change at the 4.2 ka BP termination of the Indus valley civilization and Holocene south Asian monsoon variability. Geophys Res Lett 30:1425–1428

Stuiver M, Reimer PJ, Bard E, Beck JW, Burr GS, Hughen KA, Kromer B, McCormac G, van der Plicht J, Spurk M (1998) INTCAL98 radiocarbon age calibration, 24,000–0 cal BP. Radiocarbon 40:1041–1083

Thamban M, Rao VP, Schneider RR, Grootes PM (2001) Glacial to Holocene fluctuations in hydrography and productivity along the southwestern continental margin of India. Palaeogeog Palaeoclimatol Palaeoecol 165:113–127

Tiwari M, Ramesh R, Somayajulu BLK, Jull AJT, Burr GS (2006) Paleomonsoon precipitation deduced from a sediment core from the equatorial Indian Ocean. Geo-Mar Lett 26:23–30

Tiwari M, Ramesh R, Bhushan R, Sheshshayee MS, Somayajulu BLK, Jull AJT, Burr GS (2010) Did the Indo-Asian summer monsoon decrease during the Holocene following insolation? J Quat Sci 25:1179–1188. doi:10.1002/jqs.1398

von Rad U, Schaaf M, Michels KH, Schulz H, Berger WH, Sirocko F (1999) A 5000-yr record of climate change in varved sediments from the oxygen minimum zone off Pakistan, Northeastern Arabian Sea. Quat Res 51:39–53

Yadava MG, Ramesh R (2005) Monsoon reconstruction from radiocarbon dated tropical Indian speleothems. Holocene 15:50–62

Zonneveld KAF, Ganssen G, Troelstra S, Verstheegh GJM, Visscher H (1997) Mechanisms forcing abrupt fluctuations of the Indian Ocean summer monsoon during the last deglaciation. Quat Sci Rev 16:187–201

Chapter 9
New Record of Magnetic Properties of Late Quaternary Sediments from the Eastern Arabian Sea (off Goa): Inferences on Palaeoclimate

S.K. Patil and A.D. Singh

9.1 Introduction

The environmental magnetic studies involving the mineral magnetic investigations on surface soils, lake and marine sediments are widely (and routinely) being used in palaeoclimatological and palaeooceanographic research (Maher et al. 1998; Kumar et al. 2005; Rao et al. 2008; Oldfield et al. 2009; Sangode et al. 2010; Itambi et al. 2010). Mineral magnetic parameters such as low (χlf) and high (χhf) frequency magnetic susceptibility, anhysteritic remanent magnetization (ARM), step wise forward and back field isothermal remanent magnetizations (IRM) and the ratios derived from them are proved sensitive indicators of soil/sediment magnetic mineral concentration, composition and their domain states (Oldfield 1991; Maher 1986). Further the variations in soil/sediment magnetic concentration, composition and their domain states are linked to the changes in the lithosphere, hydrosphere and atmosphere (Robinson 1986; Schmidt et al. 1999; Kumar et al. 2005). As these magnetic parameters can be rapidly measured by low cost instrumentation besides the ubiquitous nature of iron oxides (/minerals) in the soil/sediment samples, such studies have opened a new branch of geophysics, popularly known as Environmental magnetism (Thompson and Oldfield 1986).

The potential of mineral magnetic studies on marine sediment cores are proved in various aspects such as core correlation and as an indicator of time and frequency domain variations of terrigeneous sedimentation (Radhakrishnamurty et al. 1968; Robinson 1986; Bloemendal et al. 1992); for the determination of geomagnetic polarity and relative palaeointensity (Lowrie and Heller 1982); for understanding the diagenesis and authigenesis of magnetic minerals in the hemipelagic and coastal

S.K. Patil (✉)
Dr. K.S. Krishnan Geomagnetic Research Laboratory, Jhunsi, Allahabad, India
e-mail: skpatil@iigs.iigm.res.in

A.D. Singh
Department of Geology, Banaras Hindu University, Varanasi, India

R. Sinha and R. Ravindra (eds.), *Earth System Processes and Disaster Management*, Society of Earth Scientists Series 1, DOI 10.1007/978-3-642-28845-6_9,

environments (Karlin 1990; Canfield and Berner 1987; Kumar et al. 2005; Rao et al. 2008, 2010); and for the investigations dealing with the biogenic contributions of magnetic minerals to the marine sediments (Kirschvink and Chang 1984; McNeill 1990). The most important aspect of these studies is the correlation of down core variations of magnetic susceptibility and various remanence parameters of deep sea sediment cores to the palaeoclimatic tracers such as geochemical, isotopic and paleontogical records.

The sedimentary records of the continental margins provide a valuable source for gathering information on past climate and oceanographic changes at higher resolution as the margins are the sites of high sedimentation rates. Sediments over the continental margins of India are characterized by a mixture of terrigenous and biogenic origin. Biogenic sediments are predominantly autochthonous, but in the outer shelf region, they are in the relict form. The terrigenous component predominates in sediments along the continental margin. Additionally, authigenic minerals like green clays, phosphorite, dolomite etc., are also formed in shallow margins. Terrigenous sediments deposited over the continental margins are varied in size ranging from clay to sand. The composition, concentration and size of terrigenous sediments are controlled by their source rock and weathering and climatic conditions. Though all the substances in marginal sediments exhibit some magnetic behavior, ferruginous minerals/compounds (largely terrigenous in nature) are obviously the most suitable for the magnetic studies. Therefore, magnetic signatures of the sediments at large depict the environmental conditions on land influencing weathering, transportation and depositional processes responsible for variations in the magnetic mineral concentrations. Here, we present a 30 kyr record of magnetic properties of a sediment core retrieved from central part of the western continental margin of India to study the palaeoclimatic variations in the late glacial, deglacial and the Holocene periods.

9.2 Climate and Geological Setting

The climate of the Indian west coast is dominated by seasonal variations of monsoonal winds and rainfall patterns. During the summer monsoon, southwest region receives maximum precipitation causing chemical weathering of the exposed rocks along the coast producing lateritic soil. Although, laterites are common along the entire coast, in the study area of basaltic terrain i.e. central region, laterites are Fe rich having its concentration 70 times higher than the concentration in average shales (Sahasrabudhe and Deshmukh 1979). The ferruginous lateritic soil is transported and deposited to the shelf and slope of the continental margin. Winter season is comparatively drier and, thus contributing minor amount of sediments to the Arabian Sea. Mineralogical characteristics of the sediments along the margin varies from north to south depending upon the sediment input from the major rivers in north and through runoff from small mountain rivers in south.

The continental slope off Goa is covered by mixture of terrigenous and biogenic sediments. Terrigenous sediments are derived from chemical weathering and leaching of basaltic rocks of the western Ghat altered into lateritic soil and drained over the shelf up to the slope by local rivers during summer monsoon. The laterites of the central west coast of India are rich in Fe and Ti. As the sediments of the central part of western Indian margin are fed by runoff through laterites, concentration of magnetic minerals/elements is related to humid or arid phase prevailing over the coastal region. The aeolian input to terrigenous sediment of the central and south-western continental margin is found to be meager during recent (Naidu et al. 1985).

9.3 Hydrography

The hydrography of the area is driven by monsoonal wind induced seasonal circulations. The coastal current in summer season flows southward. During this season, an excess of precipitation over evaporation and heavy runoff from the Western Ghats creates low salinity plume in offshore region (Kumar et al. 2005). A southerly coastal current of summer season is replaced by a northerly current during winter season (Shetye et al. 1990). High biological productivity and poor ventilation results a well developed oxygen deficient zone along the western continental margin between 150 and 1,200 m water depths.

9.4 Core Samples and Methodology

A 4.7 m long gravity core (SK-17) was collected from central part of the western Indian margin (off Goa, Fig. 9.1) at 840 m water depth (lat. 15° 15′N; long. 72° 58′E). The sediment core is characterized by alternation of dark coloured laminated and light coloured homogenous litho-facies. Core was sub-sampled on board at 1–2 cm intervals. Previously, various micropaleontological (foraminifera and pteropods), mineralogical and geochemical analyses (Singh et al. 2006; Singh 2007; Anand et al. 2008 and unpublished data) were performed on core samples of SK-17 taken at different core-depth intervals. For magnetic measurement, an interval of 2–4 cm was taken. Samples from laminated intervals are rich in C_{org} and poor in $CaCO_3$, whereas samples from homogenous intervals are poor in C_{org} and rich in $CaCO_3$. The chronology of the core was constructed using a combination of several AMS ^{14}C dates and the down-core planktic foraminiferal $\delta^{18}O$ record (Singh et al. 2006).

For magnetic measurements air dried samples were packed into 8 cm^3 styrene pots. Low (0.46 kHz) frequency magnetic susceptibility measurements were carried out by using Bartington (MS2) susceptibility meter. The magnetic susceptibility of each sample was measured three times and the average value was presented as mass specific susceptibility in terms of 10^{-8} m^3/kg. An anhysteretic remanent

Fig. 9.1 Location of gravity sediment core SK-17

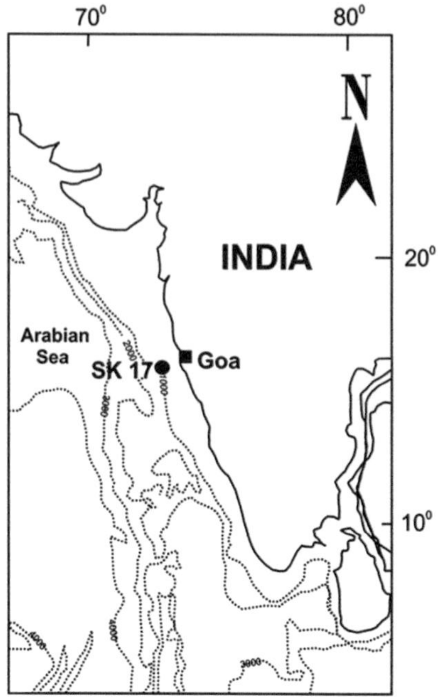

magnetization (ARM) was imparted using a 100 mT alternating field (AF) and a 0.05 mT direct current (DC) bias field. The ARM was expressed in terms of the ARM susceptibility (χARM), which was obtained by dividing the ARM intensity by the DC field strength. The χARM values are expressed in 10^{-8} m^3/kg units. Saturation Isothermal Remanent Magnetization (SIRM) was imparted using a maximum field of 1 T and the remanence was measured by Molspin spinner magnetometer. The units of SIRM are in 10^{-5} Am^2/kg. All the environmental magnetic analysis was done at Dr. K.S. Krishnan Geomagnetic Research Laboratory, Allahabad.

9.5 Results

9.5.1 Rock Magnetic Properties

Magnetic susceptibility of sediment primarily depends upon the concentration of magnetic material and their grain sizes. The low-frequency magnetic susceptibility parameter (χ_{lf}) reflects the concentration of magnetic minerals in sediment. In the studied core, the average magnetic susceptibility value is found as 13.3×10^{-8} m^3/kg ranging from 0.6×10^{-8} m^3/kg to 45.1×10^{-8} m^3/kg. The χlf in the studied core (SK-17) shows the maximum χ_{lf} values between 12.5 and 10.5 ka (Fig. 9.2).

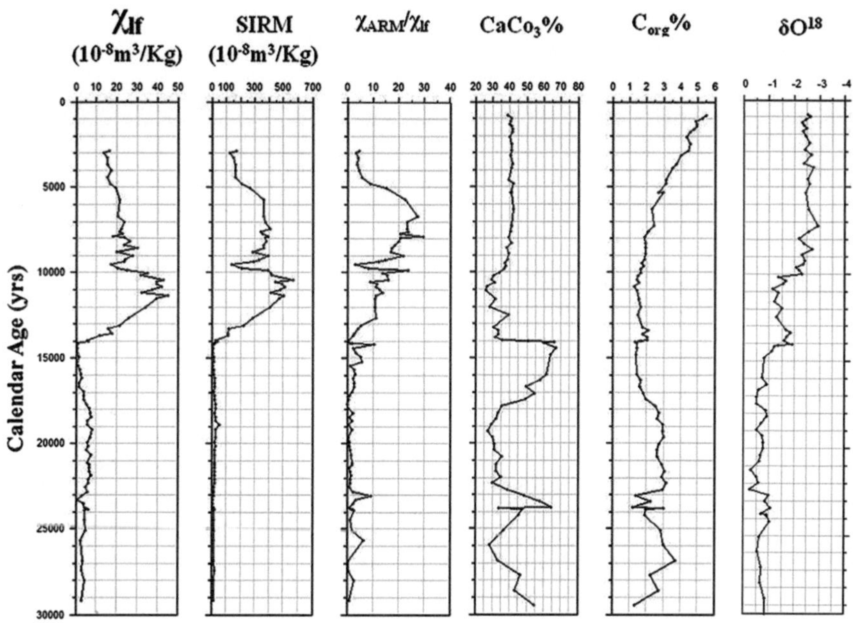

Fig. 9.2 Down core variation of δO^{18}, χlf (10^{-8} m^3/kg), SIRM (10^{-5} Am^2/kg), $\chi ARM/\chi lf$; C_{org} (%) and $CaCO_3$ (%) (Data are adapted from Singh et al. (2006) and Singh (2007))

The minimum values are recorded for the sediment older than 14 kyr BP. After a small increase between 9.5 and 8 ka (Fig. 9.2), χ_{lf} value decreases gradually in younger sediments. The SIRM reflects the concentration of ferrimagnetic materials, but unlike χ_{lf} it is not affected by relative abundances of diamagnetic and paramagnetic minerals. The SIRM value in the core ranges between 10.5 and 570.1 with maxima during 9–5.5 ka and 12.5–10 ka (Fig. 9.2). The period between 30 and 14 ka is characterized by lowest values of SIRM. The SIRM shows a prominent low value between 10–9 ka and 5–3 ka (Fig. 9.2). The χ_{ARM} (anhysteritic susceptibility) is sensitive to concentration of ferrimagnetic grains. Ratio of χ_{ARM} to χ_{lf} is used as a proxy for grain size variation of magnetic minerals (Maher 1988). Down core variation pattern of χ_{RAM}/χ_{lf} distinctly parallels to that of SIRM (Fig. 9.2).

9.5.2 Variations in Organic Carbon (C_{org}) and the Calcium Carbonate

The C_{org} content in the studied sediment core (Singh et al. 2006) shows significant increase during 28–26 28 ka, 23–18 ka and during the last 5 ka (Fig. 9.2). The minima in C_{org} are recorded during 30–29 ka, 24–23 ka and 18–15 ka (Fig. 9.2).

There is a minor increase in C_{org} content for a brief period between 14 and 13.4 ka (Fig. 9.2). The profile of total carbonate (Singh 2007) shows three intervals of maxima during 30–29 ka, 24–23 ka and 17–14 ka (Fig. 9.2). A minor increase in carbonate content is also recorded at around 13 ka. The carbonate content in sediment decreased significantly during the period from 23–17 ka, 14–13.5 ka and 12–10 ka (Fig. 9.2). Since 10 ka, there is no significant change in the carbonate content showing the moderate value of about 40%. It is noteworthy that down core variation pattern of $CaCO_3$ is opposite to that of the C_{org} through the entire core except for upper part representing the last 5 ka (Fig. 9.2).

9.6 Discussion

The magnetic parameters (χlf, SIRM, χARM/χlf) measured in core SK-17 show millennial scale variations. Broadly, the records of magnetic properties indicate major variations during three time slices: between 30 and 14 ka in the late glacial through glacial/deglacial transition period, between 10.5 and 12.5 ka in the deglacial period and between 9 and 5.5 ka in the Holocene period. Maximum value of χlf during 12.5–10.5 ka is probably linked with the high concentration of magnetic material. This χlf peak corroborates with high value of SIRM which is also an indicative of high concentration of ferrimagnetic minerals. A moderate increase of χARM/χlf (a proxy of magnetic mineral grain size) during this interval (12.5–10.5 ka) points the presence of significant proportion of fine grained material. High magnetic susceptibility can be associated with increased terrigenous supply bringing iron-bearing minerals to the core site during humid phase or through dust supply during arid phase. Minerological records of the core (unpublished data, A. D. Singh) indicate that hematite (Fe_2O_3) is the predominant magnetic mineral at this interval. The mineral Fe_2O_3 is formed in cold arid. Thus high magnetic susceptibility interval of 12.5–10.5 ka seems to be related to the arid phase. The high concentration of fine grained material (SP) further supports aeolian origin rather than fluvial which would lead to high concentration of coarse gained material. Tang et al. (2003) and Rao et al. (2008) also reported the enrichment of Fe_2O_3, Mn and χlf at YD for the sediments of the Western Pacific and eastern Arabian Sea respectively and suggested that the Fe_2O_3 is easy to enrich in the sediments during the low-temperature period. Tang et al. (2003) further suggested that the main cause for increase in susceptibility is due to the fact that the enriched Fe_2O_3, and intermingling of $MnFe_2O_4$ strengthened the area density of magnetic particles in sediments of the core and made the Fe_2O_3 distorted and recrystallized under low temperature conditions.

The period of 12.5–10.5 ka is equivalent to the northern Hemisphere cold Younger Dryas event which has been linked with a weak southwest monsoon activity. Lower part of the core representing a period between 30 and 14 ka is characterized by low values of χlf, SIRM and χARM/χlf ratio indicating low concentration of magnetic minerals through entire period. Low magnetic susceptibility of sediments for the entire period including the last glacial maximum (22–18 ka) and north Atlantic cold Heinrich equivalents (17–15 ka, 24–23 ka) that are known as arid phases (weak southwest monsoon) in the region is observed. The question arises, why magnetic susceptibility was not high during these northern Hemisphere cold phases as was observed in Younger Dryas period? There could be two plausible reasons (1) dilution of magnetic concentration by biogenic calcium carbonate and/or dissolution of fine grained iron bearing minerals. We compared our magnetic susceptibility records with the records of $CaCO_3$ and Organic Carbon contents. It is observed that $CaCO_3$ concentration was very high during 17–15 ka and 24–23 ka and organic carbon concentration was at its maximum during the last glacial maximum. This suggests that even if there was terrigenous input at core site during arid phases (17–15 ka and 24–23 ka), which probably would have been diluted by high content of $CaCO_3$. The core site is within the oxygen minimum zone, therefore oxygen deficient bottom condition developed in certain times (28–26 ka, 23–18 ka and since the last 5 ka) due to high primary productivity and high organic supply will have profound influence on dissolution/preservation conditions of biogenic and terrigenous materials. The $CaCO_3$ content profile of the core is controlled by the variation in calcareous shell preservation condition related to change in the OMZ (Singh et al. 2006; Singh 2007). It is interesting to note that C_{org} concentration was very high in sediments between 18 and 23 ka, a period when OMZ was intensified. It has been observed elsewhere that magnetic concentration can be significantly reduced by dissolution during the diagenesis (Karlin 1990). It is quite probable that there has been a major loss of iron-bearing minerals in suboxic to anoxic conditions developed during the LGM. Thus magnetic susceptibility was greatly subdued in glacial sediments by a combination of both the dilution and diagensis. The magnetic mineral concentration (χlf) in the Holocene period (the last 10 ka) does not vary much except a moderate increase during 9.5–8 ka. However, SIRM and χARM/χlf records show a major variation with their high values during 9–5.5 ka (early Holocene) and then significant decrease since 5 ka (late Holocene). Interestingly, the Holocene organic carbon profile is opposite to that of SIRM and χARM/χlf records. This suggests a diagenetic control on the magnetic susceptibility record of Holocene sediment. The southwest monsoon in the early Holocene was known to be intensified, thus high fluvial discharge bringing terrigenous material to the sea. This interpretation is also supported by our magnetic susceptible record (high SIRM values). Because of low C_{org} flux and oxygenated bottom condition prevailed at this time, finer ferrimagnetic minerals might have been better preserved. On the other hand, even if there is possibility of eolian sediments depositing at the core site in late Holocene arid period, increased flux of C_{org} and in turn oxygen depleted condition may have dissolved finer magnetic sediments resulting low magnetic susceptibility.

9.7 Conclusions

The 30 ka records of magnetic properties of the core SK-17 (off Goa) indicate major variations in the late glacial, the deglacial and the Holocene periods. Present study suggests that the magnetic susceptibility records at this core site lying within the oxygen minimum zone are governed by a combination of factors such as climatic conditions (arid/humid phases), dilution of terrigenous ferrimagnetic minerals by the biogenic $CaCO_3$ and its dissolution under high C_{org} flux and low oxygen sea bottom conditions. The period between 30 and 14 ka representing the last glacial and glacial/deglacial transition characterized by low values of χlf, SIRM and χARM/χlf, probably attributed to the dilution and diagenesis of magnetic minerals. The period between 13.5 and 10.5 ka (equivalent to the northern Hemisphere cold Younger Dryas Event) of the deglacial period characterized by high values of magnetic parameters suggest the higher influx of ferrimagnetic minerals. Study reveals that the magnetic minerals are of fine grained nature probably having aeolian origin and they were deposited during arid condition. The magnetic mineral concentration in the Holocene period (the last 10 ka) is relatively lower than that of the above deglacial period and does not vary much except a moderate increase during 8–9.5 ka. However, the early Holocene (9–5.5 ka) records show characteristically high values of SIRM and χARM/χlf parameters indicating high fluvial discharge associated with the intensification of the southwest monsoon. Whereas, the late Holocene (since 5 ka) record appear to be profoundly influenced by the dissolution of magnetic minerals due to high concentration of organic carbon.

References

Anand P, Kroon D, Singh AD, Ganeshram R, Ganssen G, Elderfield H (2008) Coupled seasurface-seawater $\delta^{18}O$ reconstructions in the Arabian Sea at the millennial scale for the last 35 ka. Paleoceanography 23:PA4207. doi:10.1029/2007PA001564

Bloemendal J, King JW, Hall FR, Doh SJ (1992) Rock-magnetism of late Neogene and Pleistocene deep-sea sediments: relationship of sediment source, diagenetic processes and sediment lithology. J Geophys Res 97:4361–4375

Canfield DE, Berner RA (1987) Dissolution and pyritization of magnetite in anoxic marine sediments. Geochim Cosmochim Acta 51:645–659

Itambi AC, von Dobeneck T, Dekkars MJ, Frederichs T (2010) Magnetic mineral inventory of equatorial Atlantic Ocean marine sediments off Senegal-glacial and interglacial contrast. Geophys J Int 183:163–177

Karlin R (1990) Magntite diagenesis in marine sediments from Oregan continental margin. J Geophys Res 95:4405–4419

Kirschvink JL, Chang SBR (1984) Ultrafine-grained magnetite in deep sea sediments: possible bacterial magnetofossils. Geology 12:559–562

Kumar AA, Rao VP, Patil SK, Kessarkar PM, Thamban M (2005) Rock magnetic records of the sediments of the eastern Arabian Sea: evidence for late Quaternary climatic change. Mar Geol 220:59–82

Lowrie W, Heller F (1982) Magnetic properties of marine limestones. Rev Geophys Space Phys 20:171–192
Maher BA (1986) Characterization of soils by mineral magnetic measurements. Phys Earth Planet Int 42:76–92
Maher BA (1988a) Magnetic properties of some synthetic sub-micron magnetites. Geophys J 94:83–96
Maher BA (1988b) Magnetic properties of modern soils and Quaternary loessic paleosols: palaeoclimatic implications. Palaeogeogr Palaeoclimatol Palaeoecol 137:25–54
McNeill DF (1990) Biogenic magnetite from surface Holocene carbonate sediments, Great Bahama Bank. J Geophys Res 95:4363–4371
Naidu AS, Mowatt TC, Somayajulu BLK, Rao KS (1985) Characteristics of clay minerals in the bed loads of major rivers in India. Milleiungen Aus dem geologisch-Palaeontogischen Institute Heft 2, University of Hamburg, Hamburg, pp 559–568
Oldfield F (1991) Environmental magnetism- a personal perspective. Quat Sci Rev 10:73–85
Oldfield F, Hao QZ, Bloemendal J, Gibbs-Eggar Z, Patil SK, Guo ZT (2009) Links between bulk sediment particle size and magnetic grain size: general observations and implications for Chinese loess studies. Sedimentology 56:2091–2106
Radhakrishnamurty C, Likhite SD, Amin BS, Somayajulu BLK (1968) Magnetic susceptibility stratigraphy in ocean sediment cores. Earth Planet Sci Lett 4:464–468
Rao VP, Kessarkar PM, Patil SK, Ahmad SM (2008) Rock magnetic and geochemical record in a core from the Eastern Arabian Sea: diagenetic and environmental implications during the late Quaternary. Palaeogeogr Palaeoclimatol Palaeoecol 270:46–52
Rao VP, Kessarkar PM, Thamban M, Patil SK (2010) Paleoclimatic and diagenetic history of the late Quaternary sediments in a core from the Southeastern Arabian Sea: geochemical and magnetic signals. J Oceanogr 66:133–146
Robinson SG (1986) The late Pleistocene palaeoclimatic record of North Atlantic deep-sea sediments revealed by mineral magnetic measurements. Phys Earth Planet Int 42:22–47
Sahasrabudhe YS, Deshmukh SS (1979) The laterites of Maharashtra state. Proceedings of the international seminar on laterisation processes. Laterisation processes, Oxford and IBH Publishing Co., Trivandrum, India, pp 209–220
Sangode SJ, Vhatkar K, Patil SK, Meshram DC, Pawar NJ, Gudadhe SS, Badekar AG, Kumaravel V (2010) Magnetic susceptibility distribution in the soils of Pune Metropolitan Region: implications to soil magnetometry of anthropogenic loading. Curr Sci 98(4):516–527
Schmidt AM, von Dobeneck T, Bleil U (1999) Magnetic characterization of Holocene sedimentation in the South Atlantic. Paleoceanography 14:465–481
Shetye SR, Gouveia AD, Pathak MC (1990) Vulnerability of the Indian coastal region to damage from sea level rise. Curr Sci 59:152–156
Singh AD (2007) Episodic preservation of pteropods in the eastern Arabian Sea: monsoonal change and OMZ intensity and aragonite compensation depth. Indian J Mar Sci 36:378–383
Singh AD, Kroon D, Ganeshram R (2006) Productivity and OMZ intensity variation in the eastern Arabian Sea at the millennial scale. J Geol Soc India 68:369–377, Special issue on Holocene Indian Monsoon
Tang X, Chen Z, Yan W, Chen M (2003) Younger Dryas and Heinrich events recorded by magnetic susceptibility of sediments from the central temperature area of Western Pacific Warm Pool. Chin Sci Bull 48:808–813
Thompson R, Oldfield F (1986) Environmental magnetism. Allen and Unwin, Boston

Chapter 10
Anthropogenic Climate Change: Observed Facts, Projected Vulnerabilities and Knowledge Gaps

Rajesh Agnihotri and Koushik Dutta

10.1 Introduction

Our planet earth is unique in terms of supporting various forms of life from ages, even though it has undergone several periods of significant environmental and climate changes from inter-annual to orbital (>10,000 years) or multi-millennial timescales. It is true that the nature of life or biodiversity has also undergone significant changes plausibly in response to large changes in the climate. For instance, during the Mesozoic era, spanning the time between the Triassic and the Cretaceous periods (~250–65 million years ago) when the earth's climate was dominated by lush green vegetation, it supported large reptiles such as dinosaurs, during which other forms of life could not sustain. According to one of the theories, sudden climate change (alternatively caused by asteroid impact and large scale volcanism) was possibly responsible for crossing the tolerance level and subsequent extinction of dinosaurs. Since the last ~4–5 ka (mid to late Holocene), variety of evidences suggest that human life has dominated the earth and various civilizations arrived and thrived for a significant period of time. Certain civilizations which left their imprints in the geological records have got extinct and one school of thought believes that possibly climate change played a significant role in their extinction like earlier. Notable examples include the civilizations of the Maya, the Indus Valley and the ancient Egypt (Hodell et al. 1995; Demenocal 2001; Cowie 2007). Along the similar lines, after the advent of industrial age (beginning in ~1850 AD in Europe), lifestyle of human beings changed significantly world wide. World

R. Agnihotri (✉)
National Physical Laboratory (Council of Scientific and Industrial Research), New Delhi 110012, India
e-mail: rajagni9@gmail.com

K. Dutta
Large Lakes Observatory, University of Minnesota-Duluth, 2205 East 5th Street, Duluth, MN 55812, USA

R. Sinha and R. Ravindra (eds.), *Earth System Processes and Disaster Management*, Society of Earth Scientists Series 1, DOI 10.1007/978-3-642-28845-6_10,

population (especially in the third world countries) also has increased in tandem with industrial activities since then. As industrial activities demand large energy production from available natural resources and emit large amount of greenhouse (heat trapping) gases in atmosphere, scientists carefully started monitoring earth's environmental variables e.g., surface air temperatures, sea surface temperatures, sea levels and humidity. They observed the rapid changes in climate variables in near tandem with the increasing levels of ambient GHGs. These observations led to the creation of new geological era called *Anthropocene*, in which most of the anthropogenic activities are thought to be governing the earth's environmental variables (Rockstrom et al. 2009). Many recent researches are aimed to estimate role of anthropogenic forcing on earth's climate. To answer this vital question a highly multi-disciplinary approach is desired as earth system science involves qualitative and quantitative assessment of almost all possible human activities and its side-effects.

Observed and projected changes in the earth climate system mainly caused by human activities commonly termed as *anthropogenic climate change*. We first review some of major observed facts in the first part of our communication followed by projected vulnerabilities of anthropogenic climate change and then we briefly discuss important knowledge gaps that are yet to be fully understood.

10.2 Observed Facts

10.2.1 Average Rise in Global Temperature

Average global temperature has increased by about 0.26°C over the twentieth century. Intergovernmental Panel for Climate Change (IPCC) comprising more than thousands of scientists have proposed with high confidence that global temperatures will continue to rise for decades to come, largely due to greenhouse gasses produced by human activities (IPCC 2007). Forecasts indicate that a possible temperature rise of ~0.7°C–2.6°C over the next century. Figure 10.1 shows global temperature anomalies since AD 1850–2009 depicting a clear and sharp rise in average global temperature. We have used here the annual Northern Hemisphere (NH) temperature anomaly data reconstructed by P.D. Jones, D.E. Parker and colleagues [http://cdiac.ornl.gov/ftp/trends/temp/jonescru/nh.dat]. It appears little unlikely to fully attribute this rising temperature trend of earth surface to greenhouse warming as some part of this rise in temperature could also be of natural origin. As earth has experienced a colder period popularly known as Little Ice Age (~1600–1650 AD; Agnihotri et al. 2002 and references therein), it could be possible that earth is still recovering from that cold period.

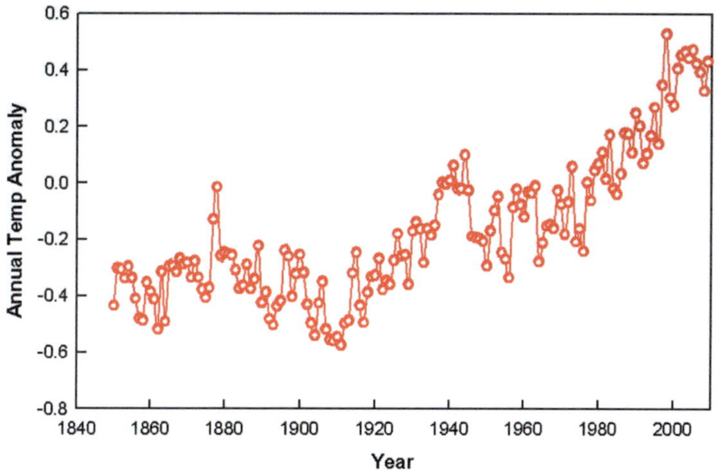

Fig. 10.1 Global annual temperature anomalies (in °C) relative to the 1961–1990 mean from Jones and Parker' data (Source: http://cdiac.ornl.gov/ftp/trends/temp/jonescru/global.txt)

10.2.2 Natural Climate Control Loosing Its Grip on Global Temperature Rise

Though strongly debated, voluminous paleo-data unarguably suggest that various manifestations of terrestrial climate vary in unison with contemporaneous solar internal variability on multi-decadal timescales (Agnihotri et al. 2002, 2011; Bond et al. 2001; Verschuren et al. 2000; Mangini et al. 2005; Wang et al. 2005). John A. Eddy made observations that cold and warm periods of northern Europe co-occurred with low and high solar activity periods respectively (Eddy 1976). Eddy's pioneering work led to many other investigations that revealed the fact global average temperatures show covariance with contemporary solar intrinsic variability (Sunspot activity). In fact, significantly cooler temperatures in Europe in the late sixteenth and early seventeenth centuries (geologically termed as Little Ice Age, LIA) contemporaneous with two major low solar activity periods i.e., the Maunder and the SpÖrer minima clearly indicated intimate linkage between terrestrial climate and external solar forcing. Recent studies dealing with annual time series of precipitation in Indian sub-continent (Bhattacharyya and Narasimha 2005), equatorial Africa (in terms of lake levels; Stager et al. 2007) and South America (in terms of stream flow major rivers; Mauas et al. 2008, 2010) and contemporaneous Sunspot activity further reinforce above hypothesis. In terms of surface temperatures also, recently Lockwood et al. (2010) showed concurrence of major anomalous central England with periods of low solar activity. In aggregate, solar intrinsic variability is probably the basic underlying forcing providing the basic ramp for natural climate variability on earth on decadal timescales (Gray et al. 2010;

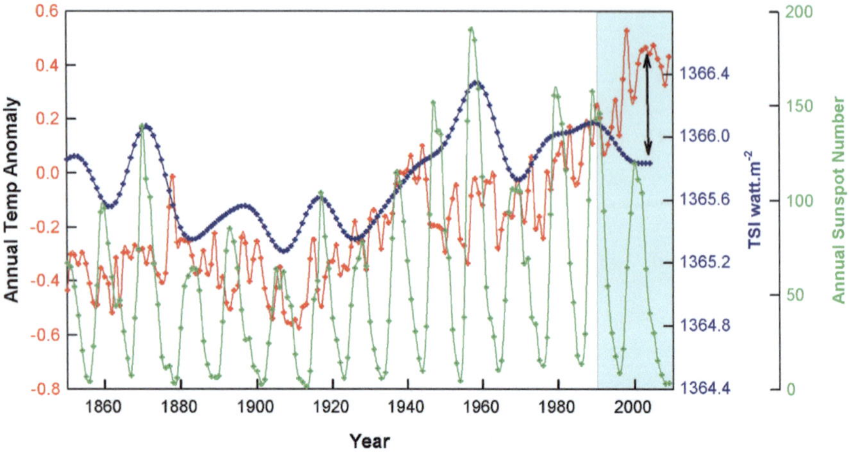

Fig. 10.2 Breaking covariance with global annual temperature anomalies (in °C) with respect to indices of solar energy output shown by *black arrow* (TSI data are adapted from Krivova et al. (2007) and data of annual sunspot number from http://www.ncdc.noaa.gov/paleo/forcing.html)

Agnihotri et al. 2011). The major concern relevant for recent few decades is that "Has solar forcing loosened its grip from terrestrial climate?" Northern hemisphere temperatures at least appear to show that since ~1975 onwards (Fig. 10.2).

10.2.3 Eurasian Snow Cover Decrease

Eurasian snow cover (ESC) extent is considered as an important determinant of various climate manifestations such as south Asian monsoon, owing to its plausible influence on land-ocean thermal contrast. H.F. Blanford was the first one to suggest summer monsoon over India and Burma (Myanmar) might be influenced by the snow cover over the Himalayas during the preceding spring season. He made the first forecast in 1881 on the hypothesis that the varying extent and thickness of snow exercise a great and prolonged influence on climate conditions and weather of the plains of northwest India (Blanford 1884). A snow cover index at the end of snow season over the Himalayas was constructed by G. Walker who obtained a correlation of −0.4 between former and the subsequent June to September rainfall over northwest India for the period 1876–1910 (Walker 1910). As outgoing longwave radiation (OLR) is increasingly getting trapped by atmosphere in response to enhancing levels of GHGs (global warming), snow cover over Himalayas is expected to decrease. Figure 10.3 shows steep decrease in ESC extent (March) since 1980 onwards. This trend is expected to enhance land-ocean thermal contrast and thereby wind intensity during SW monsoon. Using satellite observations Goes et al. (2005) showed intensity of southwest monsoon (SWM) winds appear to have intensified significantly during the last decade, enhancing upwelling induced

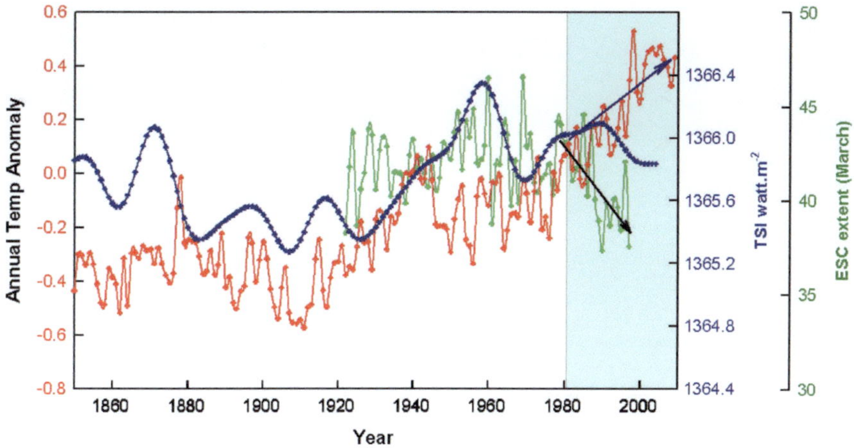

Fig. 10.3 Steep decrease observed in Eurasian snow cover (March) since AD 1980 shown by *black arrow*. Eurasian snow cover extent (SCE) derived from reconstructed daily snow depth (1922–1971) and NOAA satellite data (1972–1997). The method for reconstructing snow cover extent is described in Brown (2000) (Data source: http://nsidc.org/data/docs/noaa/g02131_brown_snow_cover/index.html)

oceanic biological productivity in the western and central parts of the Arabian Sea. This observation they interpreted as a plausible response to global warming in terms of declining ESC extent. Overall surface productivity in the northern Indian ocean has not seemed to have increased (Prakash and Ramesh 2007) leaving this question still open. Role of Eurasian snow cover extent in determining next year Indian monsoon rainfall or its oceanic manifestations is thus still not fully understood as evidenced from most of the published works.

10.2.4 South Asian Atmospheric Pollution (Aerosols)

Various modern anthropogenic activities (e.g. agricultural, biomass-fossil fuel burning and industrial emissions) have resulted in an unprecedented increase in inventory of anthropogenic aerosols in the lower atmosphere (Fig. 10.4). This is obvious with increasing haze and smog problems during winter in tropical countries affecting humans economically as well as health wise. Primary role of aerosols is to provide blockage to incoming solar radiation, hence producing a cooling effect on earth. However, there are several possible indirect effects of these suspended tiny particles in atmosphere such as effects on regional hydrological cycle, chemical state of the interacting atmosphere etc. Overall impact of these anthropogenic aerosols constitutes one of the largest sources of uncertainty in the quantification of climate change (IPCC 2007) and hence opens up several new avenues for intense and in-depth research. Qualitatively direct effects of aerosols may have been

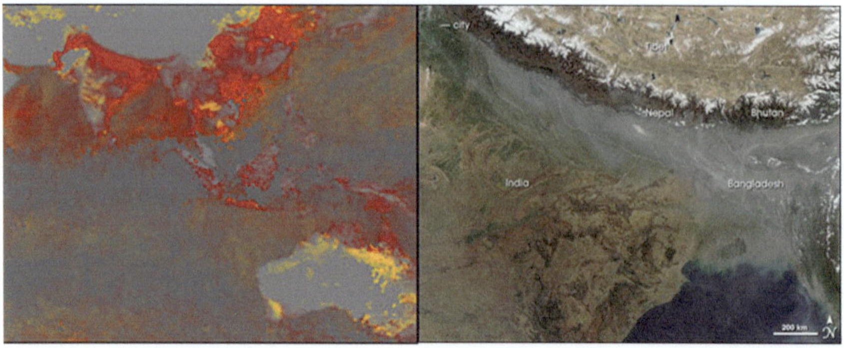

Fig. 10.4 Aerosol optical depths reported by NASA's Earth-observing 'Terra' satellite on December 8-12-2004. *Red color* tags small particles and *gold color* large particles (http://earthobservatory.nasa.gov/IOTD/). Similarly, polluted atmospheric *brown* clouds over north and northeastern India can be viewed by satellite images (Source: http://www.newscientist.com/article/dn10722-brown-haze-over-india-harming-rice-harvest.html)

understood, however, quantification of these is still underway by various research groups world wide. As far as indirect effects of aerosols are concerned, they are not completely understood yet. Ramanathan et al. (2001) has projected a weaker hydrological cycle and plausible scarcity of fresh water for South Asia, in future, as a side effect of increasing anthropogenic aerosols. In parallel, aerosols can affect the cloud-radiation feedback and the precipitation over the Indian monsoon region (Niyogi et al. 2007). New researches find that increased absorbing aerosols in the Indo-Gangetic Plain in recent decades may have led to long-term warming of the upper troposphere over northern India and the Tibetan Plateau, enhanced rainfall in northern India and the Himalayas foothill regions in the early part (May–June) of the monsoon season, followed by diminished rainfall over central and southern Indian in the later part (July–August) of the monsoon season (Lau and Kim 2010 and references therein). A recent research (Bollasina et al. 2011) suggests a plausible weakening of Asian Summer monsoon due to anthropogenic aerosols mainly emitted from European region. Soot and black carbon (BC) content of the aerosols, which is produced by partial combustion of bio-fuels, has become emerged as a prime agent having great potential to affect terrestrial climate in foreseeable future. This is because BC is the dominant absorber of visible solar radiation in the atmosphere and is expected to heat the atmosphere in a mixture with other aerosols. In fact, Ramanathan and Carmichael (2008) characterized emissions of BC (mainly from tropics) as second strongest contributor of current global warming after CO_2 emissions. Menon et al. (2002) has projected BC as a leading agent responsible for rapid melting of high altitude glaciers. Figure 10.5 shows increasing global BC emissions in atmospheric air since industrial activities commenced in Europe. Exact identification of sources and quantification of BC inventories from South Asia is pre-requisite in order to come up with feasible and affordable mitigating technologies for energy production so that current economic growth of the region can be sustained with minimal adverse climatic impacts.

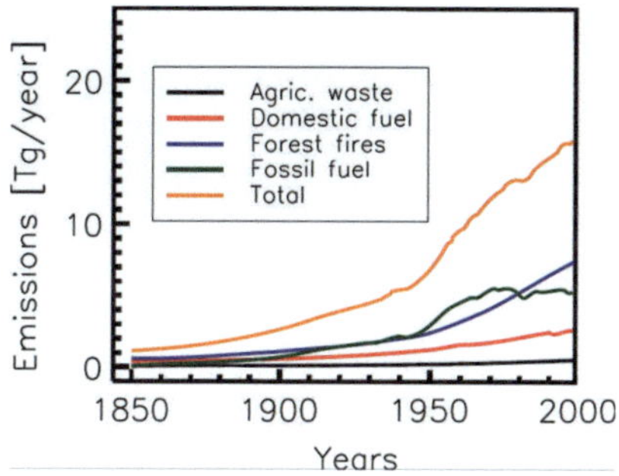

Fig. 10.5 Increasing black carbon emissions in atmospheric air since industrial revolution (~1850 AD). This figure has been provided by Rainer Lohmann (Lohmann et al. 2009)

10.2.5 Increasing Frequency of Extreme Events (Cyclones)

Global warming has also been seen as impacting weather patterns in terms of increasing in the number and severity of tropical cyclones (Webster et al. 2005). Recent researches have shown that not only are the ocean surfaces are becoming warmer, but there seems to be penetration of the human induced warming into the deeper parts of the oceans (Barnett et al. 2005). Arabian Sea has warmed by about 0.5°C during 1904–1994 (Rupa Kumar et al. 2002). Prasanna Kumar et al. (2009) explored recently the response of the Arabian Sea against global warming and its possible impact on frequency and intensity of cyclones, summer monsoon rainfall, wheat production, land vegetation cover and frequency of heat spells. They showed that AD 1995 is the demarcating year after which sea surface temperatures (SSTs) of the Arabian Sea appear to have migrated from its natural variability. As SST's of the Arabian Sea are very important determinant of regional monsoon, this departure of SSTs could manifest itself in the regional monsoonal (hydrological) cycle, impacting agriculture of the Indian sub-continent. They observed a fivefold increase in the occurrence of most intense cyclones as well as their intensity in the Arabian Sea. Winters over the northern part of the Indian landmass showed a rapid warming consistent with the increase in SST which resulted in a 16-fold decrease in the decadal wheat production after 1995. Their analyses also observed a moderate decrease of all India summer monsoon rainfall anomaly after 1995, which they linked to the decline in vegetation cover (Prasanna Kumar et al. 2009). However, as all India summer monsoon rainfall do show an epochal behavior (Agnihotri and Dutta 2003), this declining trend is part of natural variability or influenced by so called anthropogenic forcing, is a matter of debate and thus more in-depth research.

10.2.6 Melting Glaciers and Rising Sea Levels

One of the most pronounced effects of climate change has been melting of ice masses around the world. Glaciers and ice sheets are large, slow-moving assemblages of ice that cover about 10% of the world's land area and exist on every continent except Australia. They are the world's largest reservoir of fresh water, holding approximately 75% of fresh water. One of the best-documented examples of glacial retreat has been on Mount Kilimanjaro in Africa. It is the tallest peak on the continent; despite being located in the tropics it is high enough so that glacial ice has been present for at least many centuries. Of the ice cover present in 1912, 85% has disappeared and 26% of that present in 2000 now has gone (Thompson et al. 2009). If this rate of loss continues, its glaciers will likely disappear within the next decade. Similar glacial melting is occurring in Alaska, the Himalayas, and the Andes (IPCC 2001). Here it is of secondary importance that how ice is melting but the primary concern is with what rate? It is the rate which can bring plausible imbalance in global glacier mass balance. The glacier-mass balance of Greenland Ice sheet has been shown to be negative in the past few years. In contrast, South Pole (Antarctica) shows spatial heterogeneity. Recent measurements indicate that the East Antarctic ice sheet (which is much larger than the West) is gaining mass because of increased precipitation whereas west Antarctica is showing loss of ice at a quicker pace like Greenland. However, this gain in mass by the East Antarctic ice sheet is not comparable to the loss from the West Antarctica. Therefore, the mass balance of the entire Antarctic Ice Sheet also seems to be negative (Rignot 2006).

Impacts of accelerated glacier retreat are manifold. First less runoff likely to reduce capability to irrigate crops as freshwater dams and reservoirs may go dry. Water shortages could be severe in some parts of South America and Central Asia, where summertime runoff from the Andes and the Himalayas could be limited. As the volume of runoff decreases, energy production for rapidly increasing urbanization and agricultural infrastructures are likely to be stressed. This may not be immediate effect but the second impact, i.e., consequent sea level rise is a major concern in modern climate change era.

As world's ~40% human population live in coastal cities and earn their livelihood, there is an immediate concern of global warming led sea level rise especially in low lying areas like Bangladesh, Myanmar, Maldives of their very existence. Though, rates of sea level rise and its causes are topics of active debate, a recent study of Jevrejeva et al. (2009) finds most of the (~75%) of sea level rise in the Anthropocene (~1800 AD onwards) is attributable to anthropogenic forcing to a rapid increase in CO_2 and other greenhouse gases. Current estimates of global sea level rise is about 1.5–2.0 mm year^{-1}. Projections suggest that the rate of sea level rise is likely to increase during the twenty-first century, although there is considerable controversy about the likely size of the increase. This controversy is mainly due to uncertainties about the contributions from the three main processes responsible for sea level rise that are: thermal expansion, the melting of glaciers and ice caps, and the loss of ice from the Greenland and West Antarctic ice sheets. Side-effects of sea level rise are coastal erosion, wetland and coastal plain flooding, salinization of aquifers and soils, and a loss of habitats for fish, birds, and other wildlife and plants (Fig. 10.6).

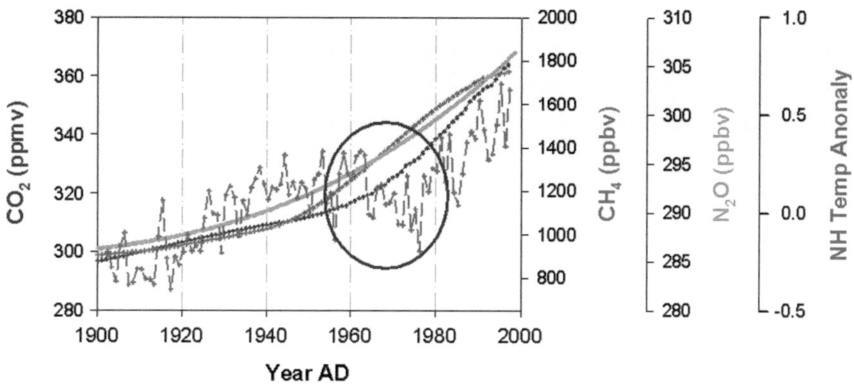

Fig. 10.6 Increasing greenhouse gases are assumed as a likely cause for increasing global air temperatures. However, the northern hemisphere temperature anomaly shows an abrupt deviation from its steadily increasing trend during 1970's (*black circle*)

10.2.7 Human Health Impacts

Global climate change may affect human respiratory systems by changing levels of air pollutants (trace gases, tropospheric Ozone, aerosols) and pollens. In addition, global warming led higher land temperatures may lead to an increase in the tropical vector-borne diseases (Rogers and Randolph 2000; Hopp and Foley 2003) thus causing concerns for human health (Sunyer and Grimalt 2006). Climate change will also affect human health somewhat less directly by affecting the environment and ecosystems in which we live in. These indirect effects will occur through insect- and rodent-transmitted diseases (e.g., West Nile virus fever; Lyme disease and Hantavirus Pulmonary Syndrome), increased smog and air pollution, waterborne and food-related illnesses and stronger UV radiation, which is a leading cause of skin cancer and cataracts. The time frame for the emergence of these health impacts can range from immediate (e.g., a cyclone/storm-related injury), to weeks or months (e.g., epidemic induced by floods, or an insect-transmitted infection such as malaria), to years and decades (e.g., UV-related malignancies such as skin cancers).

Most vulnerable regions for impacts on human health are expected to be Africa and Southeast Asia. Most of the increased burden on human death and diseases are expected from malnutrition, diarrhea, malaria, heat waves, and floods. It is not only tropical countries but highly developed ones having relatively less population and pollution may also be impacted severely by sudden change of climate. Year 2003 summer heat wave in Europe could be seen as an example, when land temperatures were ~10°C above the 30-year average. This particular heat episode killed nearly 21,000–35,000 people in countries like Belgium, France, Germany, Czech Republic, Italy, Spain, Netherlands, UK and Switzerland. France outstood in terms of death toll especially for aged people in response to this summer heat wave. Similarly in India, heat wave of 2003 at Andhra Pradesh was an unprecedented

27 days long (State of Environment Report on Andhra Pradesh 2003, http://envis-soe.ap.nic.in/pubs.html), which resulted in deaths of over 1,300 people in the state (http://www.unicef.org/media/media_6986.html).

Among increasing vector borne diseases, dengue fever, chickungunya have become most deadly as they are mosquito driven. The number of months with average temperatures higher than 18°C and the degree of urbanization were found to correlate with increasing risk of dengue fever (Wu et al. 2009). Temperature affects insect survival time and habitats as well as maturation and infective periods, and higher temperatures shorten the incubation period and viral development rate. *Ae. Aegypti*, the mosquito responsible for dengue that used to breed in small natural water bodies like tree holes or rock pools. Now, it also breeds in water that has accumulated in trash (bottles, pipes, plastics, tires). Furthermore, *Ae. Aegypti* prefer to live inside buildings rather than outside, and prefer to feed on humans instead of animals. Therefore, these mosquitoes are considered to have adapted to the urban environment which are growing with modernization and development in South Asia.

10.3 Projected Vulnerabilities

Effects of rapid industrialization, increasing agricultural activities to feed increasing population are manifold on our climate system. In an Indian context, some of the major projected vulnerabilities are listed here.

1. Steady increase in mean land temperatures may just cross tolerance level for human habitat. A warming trend has been observed along the west coast, in central India, the interior peninsula, and north-eastern India. In contrast, cooling trends have been observed in north-west India and parts of south India.
2. While all India monsoon rainfall data show majority of its inter-decadal variability is likely to be caused natural external solar intrinsic forcing (Agnihotri et al. 2011), it is also anticipated that frequency and magnitude of extreme rainfall events (or droughts) are increasing in central part of India (Goswami et al. 2006). This would result hazards related to extreme rain events such as 26 July 2005 rainfall in Mumbai, however, it is not implied that heavy rain event of Mumbai was clearly example of anthropogenic forcing as such extreme events have occurred in the past too. This scenario nonetheless poses a challenge and need to increase our preparedness for this kind of monsoon variability. In the year 2010, while major parts of India experienced normal or above-normal rainfall, plains of eastern India (e.g., eastern Uttar Pradesh and Bihar) witnessed drought type of conditions. Again notwithstanding whether this observed heterogeneity is attributable to global warming or not, in response to anthropogenic climate change, country has to increase its preparedness.
3. Global warming induced changes in key climate variables such as temperature, precipitation, and humidity, may have significant long-term implications for the quality and quantity of water. Severe changes may occur in the glacier fed river

systems owing to the accelerated melting of the Himalayan glaciers. Significant shortages in fresh water aquifers are anticipated that are dependent on water flows of large glacier fed rivers e.g. Brahmaputra and Ganga, especially during lean seasons. Water crisis is already increasing and evident in central India (Maharashtra, Madhya Pradesh) during early summer before arrival of monsoon in last decade or so, more prominently.

4. In India, significant sea level rises have been observed along the Gulf of Kutch and the coast of West Bengal (by ~0.4–2 mm $year^{-1}$). Overall, sea level rise may not appear as immediate problem compared to smaller low lying island countries, nonetheless, owing to having burden of thick economically poor population, evacuation and rehabilitation of people in case of unavoidable situation is an not an easy and affordable option.
5. In addition to sea level rise, future climate change may severely diminish present bio-diversity in coastal seas, causing food scarcity for significant coastal population. Possible increase in anthropogenic N and P owing mainly to enhanced agricultural activities on land, in coastal areas makes them prone to eutrophication. Other precarious impact of enhanced surface biological productivity in coastal areas (such as west coast of India) may enhance rates of sub-surface denitrification and in turn inventory of greenhouse gases e.g. N_2O in atmosphere (Naqvi et al. 2000). In view of this, coastal areas have to be monitored in detail using a multi-disciplinary approach to gain insights of intra and inter-annual changes for longer periods as they may reveal processes and their feedbacks that may be of significant importance in determining overall state of climate.
6. Increasing atmospheric aerosols and their soot (black carbon) content has been anticipated as the most serious concern owing to its manifold climate impacts such as net radiative forcing of atmosphere, imparting imbalance in natural hydrological cycle and their plausible role in rapid melting of glaciers. In this particular case, detailed quantitative knowledge is too limited to assess long term mitigation policies.
7. It is simple to foresee that global warming induced temperature and humidity change are going to alter the distribution of important vector species (e.g. malarial mosquitoes) and may increase the spread of such bacterial and viral diseases to new areas. Extended summer and wet periods exceeding monsoon months in presence of atmospheric aerosols have a definite effect on air-borne transmitting diseases underscoring the importance of detailed knowledge of aero-biology in the modern climate change scenario.

10.4 Important Knowledge Gaps

1. Terrestrial climate is known to vary on variety of timescales ranging from inter-annual to multi-millennial scale. Though, voluminous research has been carried out to investigate decadal to centennial scale variability of various climate manifestations, proper qualitative and quantitative identification of the former

is still not well constrained. Proper evaluation of (multi-) decadal scale climate variability is must for delineating anthropogenic forcing of climate in the recent era. For example, despite the existence of sufficient paleo-data from across the globe revealing dominance of solar intrinsic forcing in determining overall climate variability of the late-Holocene, IPCC has underestimated role of solar forcing in governing terrestrial climate. This may be owing to lack of modeling efforts which have just begun, in fact first modeling research paper incorporating solar cycles in Pacific climate has appeared recently (Meehl et al. 2009). Exact mechanist link between climate variability on earth and primary underlying natural forcing is far from understood even now. Parallel efforts are needed to determine role of underlying natural forcing of climate as efforts across the globe are underway to quantitatively determine role of anthropogenic forcing on climate.

2. Terrestrial climatic manifestation (for example monsoonal rainfall variability) is the end-product of several internal amplifiers which have to get aligned against a perturbation (internal or external) in order to produce a substantial effect in climate. However, existing knowledge of internal climate feedback processes and their relative importance is very rudimentary. How different climate drivers amplify small solar forcing on decadal scale and produced a significant and discernable change in climate is still an open question. First, response and mutual coupling processes aligning internal climate drivers have to be identified and then their roles have to be quantified to determine sensitivity of anthropogenic perturbations on climate.
3. Ocean's role in response to anthropogenic forcing is still highly uncertain. Oceans are having greater thermal inertia so likely to respond anthropogenic thermal perturbations much later than land vegetation, ice cover etc., however, since oceans store greater proportion of water and major source of water vapor that is a major greenhouse gas, any small anthropogenic perturbation capable of affecting balance between evaporation and precipitation and may get amplified in overall global warming. Another important role of oceans is its surface biological productivity which sequesters atmospheric CO_2 and transfers it to sub-subsurface levels at least extending its re-arrival in atmosphere for some hundreds to thousands of years. Therefore, oceans' carbon sequestration has been seen as a viable pathway to tackle increasing CO_2 levels. However, it is still not clear with anthropogenically enhanced ocean productivity will do the job as desired? It would be even worse if anthropogenic fertilization results in some other more greenhouse gases as by-products e.g. N_2O, which is ~270 times more potent greenhouse gas compare to CO_2. It is also not known with certainty that geo-engineering of ocean surface will trap atmosphere CO_2 for sufficient period of time, else it would be catastrophic if CO_2 gets released from oceans much quicker than anticipated. In nut-shell, proposed geo-engineering mitigating tools need much more in-depth researches before implementation.
4. This is still debatable that anthropogenic forcing is riding on the natural climate variability or former has shifted the later into another (irreversible?) climate state. Certainly concerns are logical, but from the data obtained of climate variables in

recent instrumental recording era, actual delineation and quantification of anthropogenic forcing from the natural variability are not yet achieved. More data and modeling work are required to assess future state of human influenced earth's climate.

Acknowledgements RA thanks Director, NPL New Delhi for facilities and *Council of Scientific and Industrial Research* (CSIR) for financial support.

References

Agnihotri R, Dutta K (2003) Centennial scale variations in monsoonal rainfall (Indian, east equatorial and Chinese monsoons): manifestations of solar variability. Curr Sci 85:459–463

Agnihotri R, Dutta K, Bhushan R, Somayajulu BLK (2002) Evidence for solar forcing on the Indian monsoon during the last millennium. Earth Planet Sci Lett 198:521–527. doi:10.1016/S0012-821X(02)00530-7

Agnihotri R, Dutta K, Soon W (2011) Temporal derivative of Total Solar Irradiance and anomalous Indian Summer monsoon: An empirical evidence for a sun-climate connection. Journal of Atmospheric and Solar-Terrestrial Physics 73:1980–1987

Barnett TP, Pierce DW, AchutaRao KM, Gleckler PJ, Santer BD, Gregory JM, Washington WM (2005) Penetration of human-induced warming into the world's oceans. Science 309:284–287

Bhattacharyya S, Narasimha R (2005) Wavelet cross-spectral analysis of Indian monsoon rainfall and solar activity. Geophys Res Lett 32:L05813. doi:10.1029/2004GL021044

Blanford HF (1884) On the connection of the Himalayan snowfall with dry winds and seasons of draughts in India. Proc R Soc London 37:3–22

Bollasina MA, Ming Y, Ramaswamy V (2011) Anthropogenic Aerosols and the Weakening of the South Asian Summer Monsoon. Science 334:502–505

Bond GB, Kromer B, Beer J, Muscheler R, Evans MN, Showers W, Hoffmann S, Lotti-Bond R, Hajdas I, Bonani G (2001) Persistent solar influence on North Atlantic climate during the Holocene. Science 294:2130–2136

Brown RD (2000) Northern hemisphere snow cover variability and change, 1915–1997. J Climate 13:2339–2355

Cowie J (2007) Climate change: biological and human aspects. Cambridge University Press, Cambridge

Demenocal PB (2001) Cultural responses to climate change during the late Holocene. Science 292. doi:10.1126/science.1059827 DOI:dx.doi.org

Eddy JA (1976) The Maunder minimum. Science 192. doi:10.1126/science.192.4245.1189

Goes GI, Thoppil PG, Gomes HR, Fasullo JT (2005) Warming of the Eurasian landmass is making the Arabian Sea more productive. Science 308:545–547. doi:10.1126/science.1106610

Goswami BN, Venugopal V, Sengupta D, Madhusoodanan MS, Xavier PK (2006) Increasing trend of extreme rain events over India in a warming environment. Science 314(5804):1442–1445

Gray LJ, Beer J, Geller M, Haigh JD, Lockwood M, Matthes K, Cubasch U, Fleitmann D, Harrison G, Hood L, Luterbacher J, Meehl GA, Shindell D, van Geel B, White W (2010) Solar influences on climate. Rev Geophys 48, RG4001, 53. doi: 10.1029/2009RG000282

Hodell DA, Curtis J, Brenner M (1995) Possible role of climate in the collapse of Classic Maya civilization. Nature 375:391–394. doi:10.1038/375391a0

Hopp MJ, Foley JA (2003) Worldwide fluctuations in dengue fever cases related to climate variability. Climate Res 25:85–94

IPCC (2001) "Graph of 20 glaciers in retreat worldwide". Climate change 2001 (Working group I: the scientific basis). http://www.grida.no/climate/ipcc_tar/wg1/fig2 18.htm

IPCC (2007) Climate change 2007 – the physical science basis. Cambridge University Press, Cambridge
Jevrejeva S, Grinsted A, Moore JC (2009) Anthropogenic forcing dominates sea level rise since 1850. Geophys Res Lett 36:L20706. doi:10.1029/2009GL040216
Krivova NA, Balmaceda L, Solanki SK (2007) Reconstruction of solar total irradiance since 1700 from the surface magnetic flux. Astron Astrophys 467:335–346
Kumar SP, Roshin RP, Narvekar J, Dinesh Kumar PK, Vivekanandan E (2009) Response of the Arabian Sea to global warming and associated regional climate shift. Mar Environ Res 68:217–222. doi:10.1016/j.marenvres.2009.06.010
Lau WKM, Kim KM (2010) Fingerprinting the impacts of aerosols on long term trends of the Indian summer monsoon regional rainfall. Geophys Res Lett 37:L16705. doi:10.1029/2010GL043255
Lockwood M, Harrison RG, Woolings T, Solanki SK (2010) Are cold winters in Europe associated with low solar activity? Environ Res Lett 5. doi:10.1088/1748-9326/5/2/02001
Lohmann R, Bollinger K, Cantwell M, Feichter J, Fischer-Bruns I, Zabel M (2009) Fluxes of soot black carbon to South Atlantic sediments. Global Biogeochem Cycles 23:GB1015. doi:10.1029/2008GB003253
Mangini A, Spotl C, Verdes P (2005) Reconstruction of temperature in the Central Alps during the past 2000 yr from a $\delta^{18}O$ stalagmite record. Earth Planet Sci Lett 235:741–751
Mauas PJD, Flamenco E, Buccino AP (2008) Solar Forcing of the stream flow of a continental scale South American river. Phys Rev Lett 101(168501):1–4
Mauas PJD, Buccino AP, Flamenco E (2010) Long-term solar activity influences on South American rivers. J Atmos Sol Terr Phys 73. doi:10.1016/j.jastp. 2010.02.019
Meehl GA, Arblaster JM, Matthes K, Sassi F, van Loon H (2009) Amplifying the Pacific climate system response to a small 11-year solar cycle forcing. Science 325:1114–1118
Menon S, Hansen J, Nazarenko L (2002) Climate effects of black carbon aerosols in China and India. Science 297:2250–2253
Naqvi SWA, Jayakumar DA, Narvekar PV, Naik H, Sarma VVSS, D'Souza W, Joseph S, George MD (2000) Increased marine production of N_2O due to intensifying anoxia on the Indian continental shelf. Nature 408:346–349
Niyogi D, Chang H-I, Chen F, Gu L, Kumar A, Menon S, Pielke RA Sr (2007) Potential impacts of aerosol–land–atmosphere interactions on the Indian monsoonal rainfall characteristics. Nat Hazards 42. doi:10.1007/s11069-006-9085-y
Prakash S, Ramesh R (2007) Is the Arabian Sea getting more productive? Curr Sci 92:667–671
Ramanathan V, Carmichael G (2008) Global and regional climate changes due to black carbon. Nat Geosci 1:221–227
Ramanathan V, Crutzen PJ, Kiehl JT, Rosenfeld D (2001) Aerosols, climate, and the hydrological cycle. Science 294:2119–2124
Rignot E (2006) Changes in ice dynamics and mass balance of the Antarctic ice sheet. Philos Trans R Soc 364:1637–1655. doi:10.1098/rsta.2006.1793
Rockström J et al (2009) A safe operating space for humanity. Nature 461:472–475. doi:10.1038/461472a
Rogers D, Randolph SE (2000) The global spread of malaria in a future warmer world. Science 289:1763–1766
Rupa Kumar R, Krishna Kumar K, Ashrit RG, Patwardhan SK, Pant GB (2002) Climate change in India: observations and model projections. In: Shukla PR, Sharma SK, Venkata Ramana P (eds) Climate change and India: issues, concerns and opportunities. Tata McGraw-Hill Publishing Company Limited, New Delhi
Stager JC, Ruzmaikin A, Conway D, Verburg P, Mason PJ (2007) Sunspots, El Niño, and the levels of Lake Victoria, East Africa. J Geophys Res 112:D15106. doi:10.1029/2006JD008362
Sunyer J, Grimalt J (2006) Global climate change, widening health inequalities, and epidemiology. Int J Epidemiol 35:213–216. doi:10.1093/ije/dyl025

Thompson LG, Brechera HH, Mosley-Thompson E, Hardyd DR, Mark BG (2009) Glacier loss on Kilimanjaro continues unabated. PNAS 106:19770–19775. doi:10.1073.pnas.0906029106
Verschuren D, Laird KR, Cumming BF (2000) Rainfall and drought in equatorial East Africa during the past 1,100 years. Nature 403:410–413
Walker GT (1910) Correlations in seasonal variations of weather, II. Mem India Meteorol Dept 21:22–45
Wang Y, Cheng H, Edwards RL, He Y, Kong X, An Z, Wu J, Kelly MJ, Dykoski CA, Li X (2005) The Holocene Asian monsoon: links to solar changes and North Atlantic climate. Science 308:854–857
Webster PJ, Holland GJ, Curry JA, Chang HR (2005) Changes in tropical cyclone number, duration, and intensity in a warming environment. Science 309:1844–1846
Wu P et al (2009) Higher temperature and urbanization affect the spatial patterns of dengue fever transmission in subtropical Taiwan. Sci Total Environ 407:2224–2233

Section III
Natural Disasters

Chapter 11
Landslide Hazard Analysis and Management: A Case Study from Nainital, India

D. Chakraborty and R. Anbalagan

11.1 Introduction

In a fragile mountain ecosystem like the Himalaya, landslide is a common mass wasting phenomenon caused by both natural and/or anthropogenic factors. Study of slope failures, particularly their hazard probability and damage potential, becomes important as human settlements and other infrastructures may get affected by such phenomenon. After assessing the status of stability of potentially unstable slopes, it is equally important to find out suitable remedial measure for those slopes so as to mitigate the risk caused by landslide occurrences.

In this context, Nainital, one of the famous hill resorts of Uttarakhand with its picturesque hill slopes, is worth mentioning, as it represents a classic example of a hill town, where disproportionate urbanization has resulted in major instability problems of the hill slopes. The Nainital town (Latitude: N 29° 22′–N 29° 24′ and Longitude: E 79° 26′–E 79° 28′) is located in Kumaun Lesser Himalaya and falls in Survey of India Toposheet No. 53 O/7. The town is situated about 35 km north of Kathgodam, which is the nearest rail head. The Kailakhan area falls on the south eastern (SE) corner of Nainital township. The Naini Lake represents a saucer shaped depression and is surrounded by Lesser Himalayan hills from all sides (Fig. 11.1). The Balia stream, the only outlet of Lake Nainital, emerges from this lake towards SSE direction and passes through Kailakhan area, before meeting Golariver near Kathgodam. The Kailakhan region thus forms the initial part of the Balia valley, where sharp stream bends can be observed causing toe erosion. Hill slopes along these bends are characterized by weak rocks that form steep side slopes of the Balia ravine. Moreover, Nainital and its surrounding areas, including

D. Chakraborty
Geological Survey of India, NER, Shillong, India

R. Anbalagan (✉)
Department of Earth Sciences, IIT, Roorkee, India
e-mail: anbautr@gmail.com

R. Sinha and R. Ravindra (eds.), *Earth System Processes and Disaster Management*, Society of Earth Scientists Series 1, DOI 10.1007/978-3-642-28845-6_11,

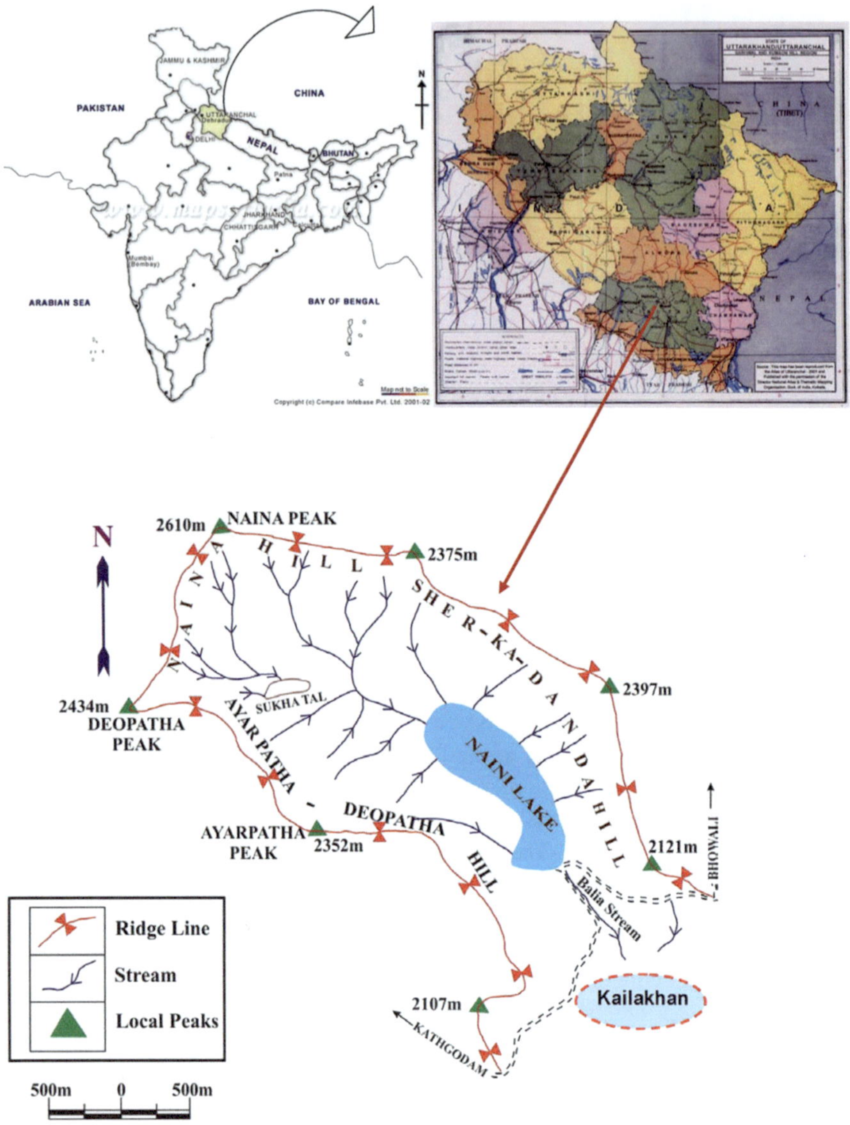

Fig. 11.1 Location of Kailakhan area within Nainital town limit

Kailakhan, receive high annual precipitation of more than 150 cm. These factors culminate to make Kailakhan a potential area for hill instability. In fact, landslide scars are clearly visible on many slopes from the distantly located Nainital–Kathgodam and Nainital–Bhowali motor roads. Though there are many slides in this area, the one, which is active, conspicuous and presently threatening the thickly populated human settlements above its crown area, has been selected for detailed slope stability study and remedial measures, has been accordingly suggested.

11.2 Geological Setting

Geologically, Nainital area is represented by rocks of Infra-Krol, Krol and Tal Formations, all belonging to Mussoorie Group (Proterozoic age). North of Kathgodam, the Outer Himalayan rocks comprising Lower Siwalik Formation (Cenozoic age) of Siwalik Group are exposed, which comprises sandstone and shale with alternating bands of mudstone. This sequence continues further north up to Jeolikot, located approximately 16 km south of Nainital, where Lesser Himalayan rocks are brought into juxtaposition with Lower Siwalik rocks by Main Boundary Thrust. After Jeolikot, rocks of Mussoorie Group belonging to Lesser Himalayan sequence, continues up to Nainital and further beyond. It consists of Blaini, Krol and Tal Formations. Named after Kailakhan area, Kailakhan member of Blaini Formation is also called Infra-Krol (Valdiya 1980, 1988) as it is stratigraphically located immediately below Krol Formation. Rocks in Kailakhan area are dominantly made up of olive-green, grey, black or even red colored finely banded shale and slate with fine grained muddy sandstone. Shales and slates show well developed foliation planes.

11.3 Site Observations

Urbanization in Kailakhan area is more concentrated by the sides of Nainital–Kathgodam road, in comparison to Nainital–Bhowali road. The general slope inclination from Nainital–Kathgodam road towards Balia stream is gentle (20°–25°) in E-ESE direction. Just adjacent to the road level, looking down, these gentle slopes show different levels where urbanization is conspicuous, till steep valley slopes (50°–60°) are encountered, close to Balia stream. Several first order streams flow through these steep slopes further down. The urbanized upper portion of the slope close to Nainital–Kathgodam road is made of moderately consolidated debris derived from higher levels. Rocks exposed further down, in the steep slopes, are mainly Infra-Krolshales and slates with minor phyllites with foliation plane as the dominant geological discontinuity having dip amount of 30°–35° towards N 320°–330° (Figs. 11.2 and 11.3). They are also affected by closely spaced single set of joint (70°–85°/N 100°–120°). Random joints are also observed at places. Close to Balia stream level, loose and unconsolidated debris are observed all throughout the right bank, which generally consists of silty sand to gravel size fractions. This debris is essentially derived from the recent slide activities along the steep slopes of Balia ravine. Almost similar kind of slope material is observed on the left bank of Balia stream. Here past landslide scars are also clearly visible. However, urbanization on left bank is practically nil as compared to right bank. Hence, the present study is mainly confined to stability analysis of right bank.

The landslide, studied in the present work (Fig. 11.2), is located on an ESE facing slope, almost 100 m from road level on a horizontal stretch, but the drop in elevation from road to its crown portion is about 40 m. The slope in between is very

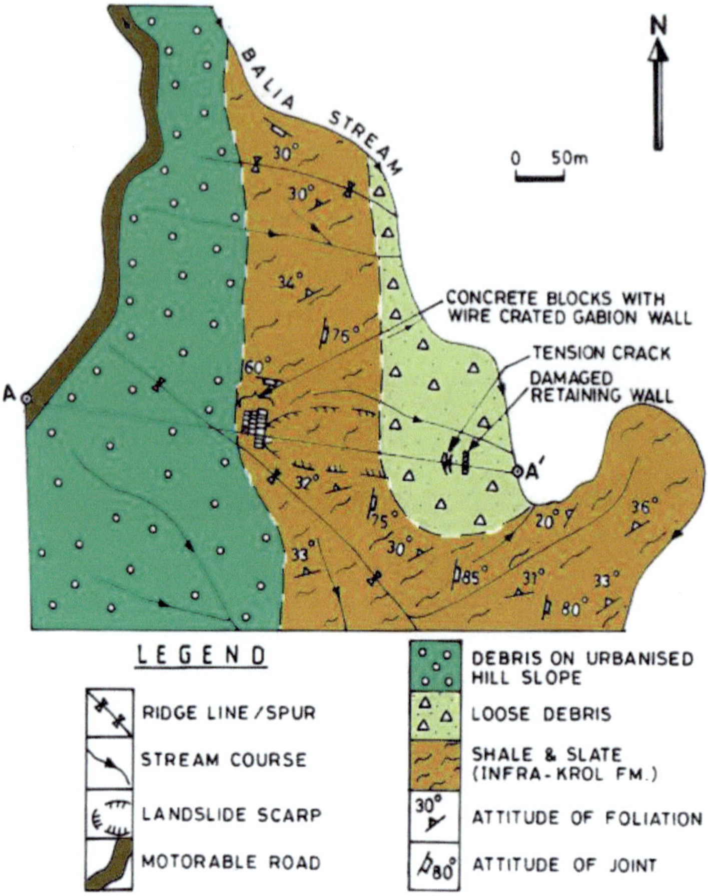

Fig. 11.2 Geological map of Kailakhan landslide area

gentle where urban structures are concentrated, as indicated in Fig. 11.3. The slided materials get accumulated at the base in the form of a fan. The maximum width of the slide is of the order of 70–80 m and the slope segment from the crown of slide to stream level shows a height of about 200 m. At the crown portion of the slide, two layers of concrete blocks can be seen with a layer of wire-crated gabion wall in between them. This is followed by a layer of highly weathered dark red shale. Rock slope comprised of Infra-Krol slates and phyllites continues further down for about an 80–100 m at average slope angle of the order of 70°, forming a scarp face. On this slope, foliations and joint planes can be prominently seen (Fig. 11.3). Foliations are closely spaced (10–15 cm) and dip into the hill at gentle angles (18°–20°). Steeply dipping joint planes have same inclination as that of rock slope face. Joint planes are persistent, weathered and show indications of dilation close to slope face. These are moderately closely spaced with average spacing of about 2–5 m and in most cases, they show persistence of the order of 10 m. Intersection of joint planes

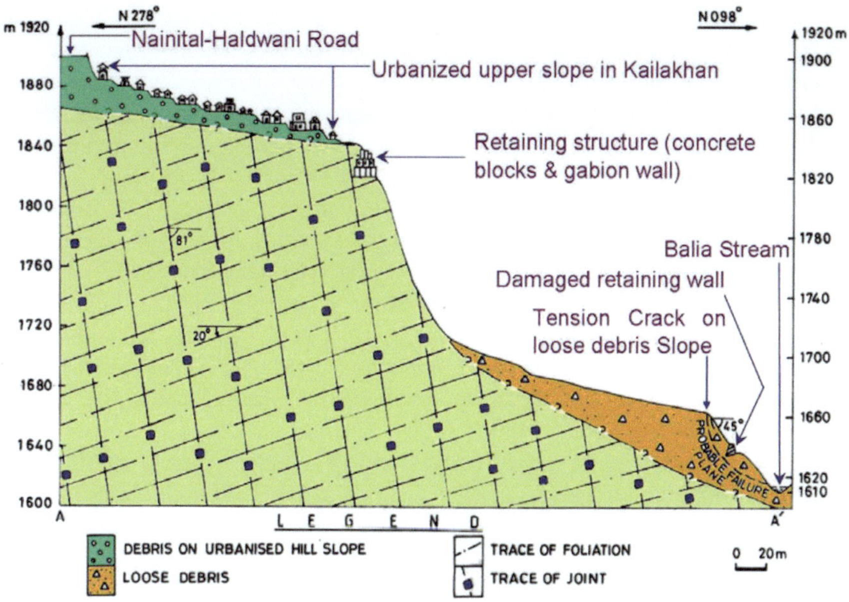

Fig. 11.3 Geological section through slide zone showing the pattern of discontinuity

with foliation plane thus forms tall and slender rock blocks. The scarp face ends with a debris cone at its base, which conceals further rock exposures down below. From this point onwards, the slope material consists of loose, unconsolidated debris with dominant coarser fragments, up to stream level. Initial 100 m of this lower part of slope segment is gentle with average slope angle varying between 15° and 20° with very minor slope breaks. This is followed by a sudden change in slope gradient, which increases to 45–50° and continues up to stream level over a height of about 50–60 m. Initial signs of slope instability in the form of tension cracks can be observed in this part of the slope where the slope gradient suddenly increases. Further 20–25 m down slope from lower slope break (Fig. 11.3), a badly damaged retaining wall is also seen, indicating unstable slope conditions.

11.4 Causes of Landslide

Anbalagan (1993) identified toe erosion by the Balia stream on both the banks as the immediate cause for repetitive slumping process in Kailakhan. The striking evidence for this observation is that the initial 500 m of Balia stream, after emerging from Nainital Lake is lined on both sides and in this segment neither of the banks show sign of slope instability. As concrete lining of the Balia stream has not progressed further, the stream banks start showing instability due to toe erosion. Further downstream of the concrete lining, the Balia stream travels roughly for

700 m with a drop of 250 m in elevation, thus following an average stream gradient of 20°. However, in this 700 m stretch the stream bed shows sudden change in bed gradient in form of small waterfalls and also it accommodates two acute sharp bends (Fig. 11.2) along stream course. The study confirms that landslide activity is more conspicuous in these sharp stream bends. Hence, it can be inferred that the weak rock type of Infra-Krol characterized by extensive presence of foliation planes and steep valley dipping joints, is mainly responsible for extensive weathering effects along these geological discontinuities. Further, the location of the study area in a high annual precipitation zone with steep stream bed gradients and steep side slopes of the stream contributed to instability of the Balia stream banks in Kailakhan area. The direct cause obviously is the depleting toe support due to toe-cutting by Balia stream, which explains the reason of repetitive slumping process of these slopes close to stream level. With this background on geology of slide zone and extent of urbanization above the slide zone, slope stability analysis of this landslide has been carried out to identify suitable remedial measures, in order to avoid future disasters.

11.5 Mechanism of Failure

Looking from crown portion of the slide zone towards the stream level, two distinct zones of instability can be identified. First is the scarp face where Infra-Krolshales, slates and phyllites are exposed. It is quite clear from the geological section (Fig. 11.3) that loose debris accumulated at its base must have been derived from this scarp face, due to past slope failures. The next part of stability analysis is identification of type of failure. Taking a closer look at the type of slope material, the first inference that can be drawn is that the steep rock slope is traversed dominantly by foliation and steep joints. Moreover close to slope face, rocks are also highly weathered. As mentioned earlier, the joint planes are generally steeper than the slope or have same dip amount as that of the inclination of the slope. Even at lower portion of this scarp face, joints are nearly vertical. This observation is sufficient to infer that there is no 'day lighting' of the discontinuity on the slope face and as such it rules out possibility of translational failure along the persistent joint planes. There are though some random joints, they are not considered in the analysis.

As observed at the site, the joint planes are persistent, closely spaced and dips steep or parallel to slope face. The foliation planes in this case acts as basal plane of separation. At the same time, close to slope face joint planes show tendency to open up. During rains, water must have filled these open spaces and give an outward lateral push to rock mass. Similar effect is also possible during extreme winter days due to repeated freezing and thawing effects, which can also lead to further dilatancy of the existing joints. Hence, the slender blocks most probably had shown a tendency of block rotation and ultimately toppled into the valley. In order to confirm this theory, attitudes of joint planes and foliations were observed and their poles are plotted in a stereonet. After delineating the zones of toppling failure (Anbalagan et al. 2007) for this scarp face (Fig. 11.4), it is observed that

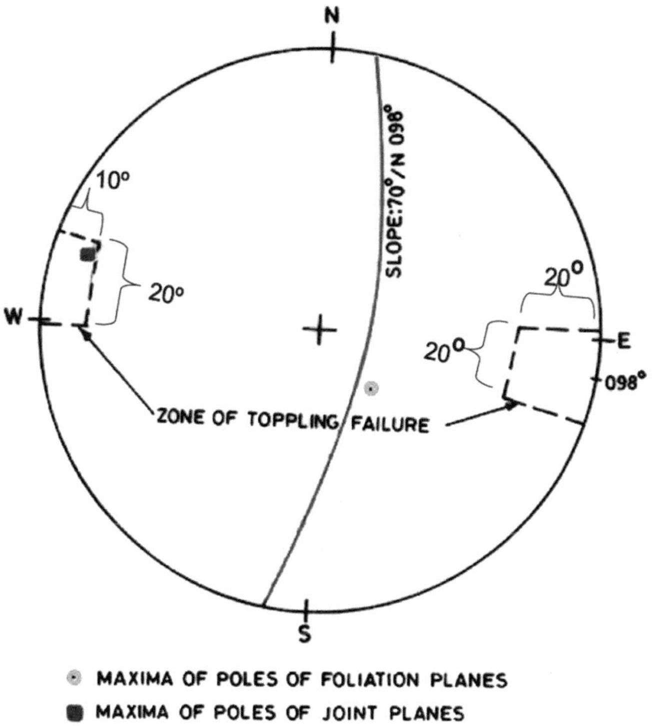

Fig. 11.4 Stereoplot of poles of geological discontinuities from Kailakhan slide zone

maxima of pole concentration of joint planes falls within one such zone, thus validating probability of toppling failure.

Another sign of instability was observed on the loose debris further down near stream level, where the average slope angle is steep (Fig. 11.3). In this segment, though thickness of debris cover could not be exactly deciphered during the study due to lack of any drilling or geophysical data, yet visual estimate indicated that it was more than 30 m. Given this depth of debris cover and slope inclination, rotational failure is the most possible mode of failure. Circular failure analysis was carried out for this part of slope segment using Circular Failure Chart (CFC) method (Hoek and Bray 1981) and by computer program SARC (Singh and Goel 2002).

11.6 Slope Stability Analysis

For stability analysis of circular mode of failure, in both the methods mentioned above, the basic input parameters such as density, shear strength parameters of the slope material, slope height and soil moisture content are required. Two samples of

Table 11.1 Details of input parameters considered for slope stability analysis

1. Height of slope (H)	60 m – from geological section (Fig. 11.3)
2. Slope angle	45° – from geological section (Fig. 11.3)
3. Type of debris (slope material)	Poorly graded gravel – GP (USCS)
4. Moisture content	<5% – dry slope condition
5. Unit weight (γ) of slope material	16.38 KN/m^3
6. Cohesion (c) of slope material	9.8 kPa – from direct shear test
7. Friction angle (Φ)	40° – from direct shear test

debris were collected from two different heights on slope for analysis and details of input parameters are listed below (Table 11.1).

11.6.1 Stability Analysis Using CFC Method

Based on the given input, the Factor of Safety as calculated for different possible slope saturation conditions are tabulated below. For dry condition Circular Failure Chart No. 1 was used. For 25% and 50% saturation, chart No. 2 and 3 were respectively used. More than 50% saturation is usually not expected as it is a hill slope composed of debris dominantly well graded gravels which is considered as a free draining material. Dimensionless ratio ($c/\gamma.H.\tan\Phi$) applicable for all the three charts is 0.024. Different values of F obtained with increasing slope saturation are indicated in Table 11.2.

11.6.2 Stability Analysis Using SARC Program

This computer program (Singh and Goel 2002) is essentially based on Bishop's method of slices (1955). The program hypothetically segments the slope into vertical strips of almost equal thickness and calculates the total force exerted by individual segments towards the toe. Failure condition is established when the cumulative force of these slope segments exceeds the mobilized shear strength along most critical plane of weakness. The failure surface in this approach is supposed to pass through the toe of the slope. Accordingly, this program seeks various combinations of failure surfaces based on the relative position of tension crack behind the crown (on top) and the exit points at the toe of slope and finally selects one combination which gives lowest F value, i.e. instability condition along the most critical failure surface. So, the SARC program helps to draw the possible failure surface as it gives the radius of curvature and coordinates of centre of rotation with respect to slope face and this is represented in geological section (Fig. 11.3). The program is executed considering existing slope saturation, i.e. 0% (dry slope) and final value of F is 1.25.

Table 11.2 F values of the slope for different slope saturation conditions

Factor of safety (F) as per the 'X' and 'Y' intercept of the chart	Slope saturation condition		
	0%, i.e. completely dry (chart no. 1)	25% (chart no. 2)	50% (chart no. 3)
Y axis intercept (F_1)	1.2	1.08	0.95
X axis intercept (F_2)	1.25	1.11	0.99
$F_{avg} = (F_1 + F_2)/2$	1.23	1.1	0.97

11.7 Discussion

Both the methods of analysis indicate a stable slope under dry condition. But from CFC method it can be seen that as the slope saturation increases, the F value goes on decreasing and the slope finally becomes unstable at 50% saturation. From these observations it can be inferred that under the condition of sudden and heavy downpour the slope may become unstable with increasing probability of failure. This inference also gets support from the ground observation that the slope shows repetitive sign of instabilities particularly during monsoon period.

11.7.1 Remedial Measures

In order to stabilize the unstable slopes, Nainital Lake Development Authority (NLDA) had already implemented some slope stabilization measures including concrete lined drains running through slide zone and constructing a series of retaining walls on the steep slope segment of the loose debris zone at different heights in staggered pattern. However it was found that these retaining walls developed several cracks and got more and more damaged every year particularly during monsoons. Consequently, based on the site observations and on stability analysis, suitable remedial measures have been identified separately for both the segments of instability within the slide zone.

11.7.2 Stabilization of Rock Slope Showing Toppling Failure Potential

In view of steepness of upper slope, the execution of control measures may be a difficult process. However, taking into consideration the size of unstable rock blocks, the only way to stabilize the slope will be by grouted rock anchors. Since the rock slopes are controlled by steeply dipping joints causing topping failure, the anchors can be driven at an angle perpendicular to the slope. The direction of providing anchors has been worked out (55° towards N 302°) after plotting the slope, joints and foliation planes on a stereo-net (Fig. 11.5) as the anchor is

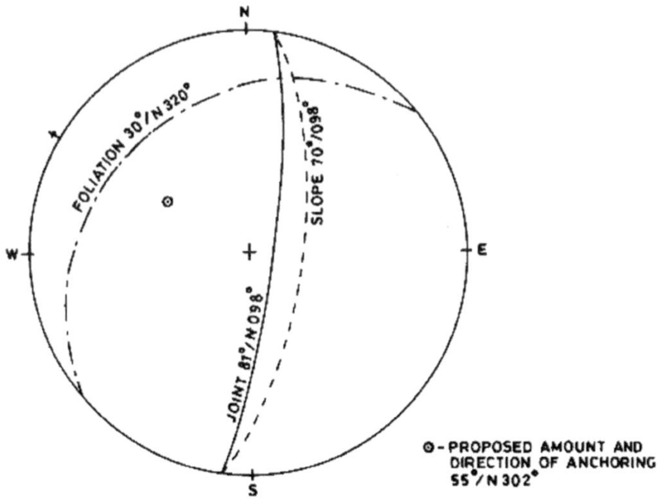

Fig. 11.5 Stereoplot showing the amount and direction of anchoring for treatment of rock face on upper slope

supposed to intersect the discontinuities at high angle (preferably orthogonal). The anchors may be driven to a length of 10–15 m at spacing of 2–3 m in staggered pattern. In addition, the rock slope should be shotcreted on a polymer wire mesh leaving suitable space for subsurface drainage. These measures will stitch the dilatant joints and effectively increase the shear strength of rock mass as a whole. However, the permanent measure would be to flatten the slope to an angle of 55°–60° as toppling failure reduces effectively when the slope inclination is reduced. The other measures as outlined above will help to stabilize the slope in a more effective way.

11.7.3 Stabilization of Steep Valley Slopes Comprising of Loose Debris Close to Balia Stream Level

In view of presence of steep slopes consisting of loose debris close to stream level and the toe erosion being an important factor in inducing the instability, the remedial measures should adequately take into consideration these factors. The design of retaining walls should effectively prevent the side slope erosion. Since this has close connection with the stream bed erosion, efforts should be focused to further arrest the bed erosion.

The first measure is intended for side slope protection by reducing the extent of toe erosion, which can be done by providing two layers of concrete cubes, placed one on top of the other, all throughout the unlined stretch of the Balia Stream (Fig. 11.6). Dimension of bottom row cube will be 1.5 m × 1 m × 1 m while that

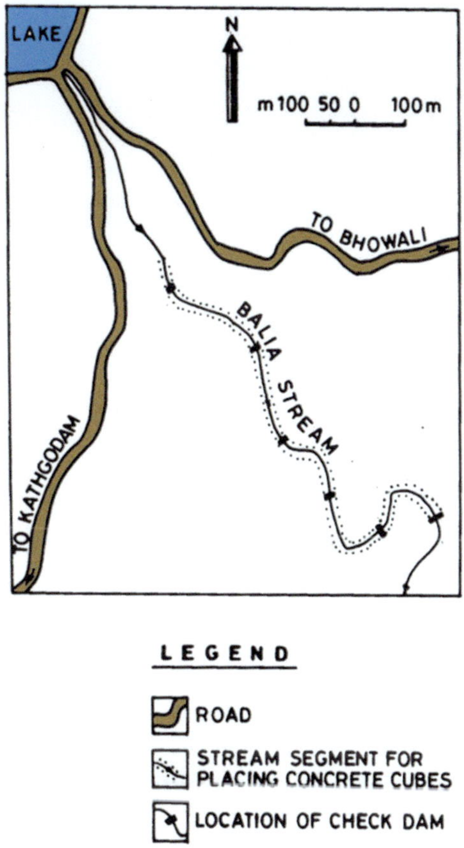

Fig. 11.6 Schematic diagram showing the stretch of Balia stream in Kailakhan area where check dams and concrete cubes are to be placed

of top row will be 1 m × 1 m × 1 m (Fig. 11.7). The bottom cube row will be placed in such a way that a gap of 10–15 cm is left in between them. This space will provide adequate drainage for subsurface water from the slope behind the cubes. These cubes will act as a toe wall for steep valley slopes and hence it will be useful to reduce toe erosion of side slopes. In this context, if any slope failure happens, that will be very shallow in nature and ultimately lead to flattening of slope angle. For effective stabilization of the debris slopes, it is suggested to provide polymer geo-grid layers over the slope and secure the layers properly to the ground. Once the concrete cubes, which are heavy structures, are placed on the stream bed just adjoining the slopes, they will not allow side slope erosion when stream water rises during monsoon. For better stability, the concrete cubes can be placed at a depth of approximately 30 cm below the stream bed (Fig. 11.7).

Since the slope consists of loose debris, bio-technical treatment may also serve as a long term measure of slope stabilization along with those mentioned so far. Fast growing grass seeds can be sprayed over the slope to render more stability to the slope. Plant species, which have deeply spread root system effectively binds the loose soil particles and indirectly increases their shear strength. Given the annual

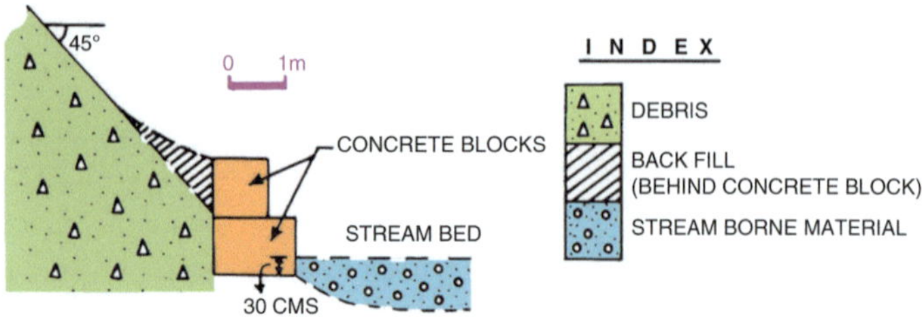

Fig. 11.7 Sectional sketch of concrete cubes, acting as toe wall for protecting side slopes of Balia Stream in Kailakhan area

precipitation rate of the area, species which have a higher transpiration rate will help more effectively to keep the slope dry.

11.7.4 Prevention of Stream Bed Erosion by Check Dams

The stream bed erosion can be effectively checked by reducing the velocity of stream water flowing through the constricted part of the Balia valley. For this purpose, taking into account general site conditions, a suitable design of check dam has been presented (Fig. 11.8). The check dam consists of handpicked boulder and rubble-fill material placed within a reinforced frame of welded G.I. pipes or steel pipes of 100 mm size, for which the dam acts as a permeable barrier and allows water to flow smoothly through it, thus reducing its velocity and subsequently preventing river bed erosion. The height of the check dam shall be 3–5 m from streambed on upstream side (Fig. 11.8) with an upstream facing gradient of 3:1. Being a flexible type of structure with low height, these dams will impart very less load to the foundation and as such can be constructed over stream bed without any foundation treatment. However, to withstand the hydraulic forces during floods, these check dams should be anchored up to 3 m, at upstream side within stream bed by making drill holes. These holes should be properly grouted, so that the structure does not get displaced from its position when horizontal forces act over it. Similarly, at the toe portion grouted anchors should be placed up to 2 m depth. From the top of the dam, looking downstream the gradient of dam should be made flatter (1:3). This is required keeping the possibility of sudden flooding in Balia Stream, when water will hit the dam structure with tremendous horizontal thrust and there is always a possibility that water may topple the dam, given its low height. In this situation a flatter downstream gradient of the dam will ensure that there will be least effect of scouring at the toe portion. As an extra precautionary measure, a 0.5 m 'vertical notch' has been provided in the design of toe portion (L-section of Fig. 11.8) for creating a 'mini stilling basin' like situation which will render spilled flood water with negligible velocity before leaving the dam structure at the toe area. The location of check dams has been suggested keeping in mind the sharp stream

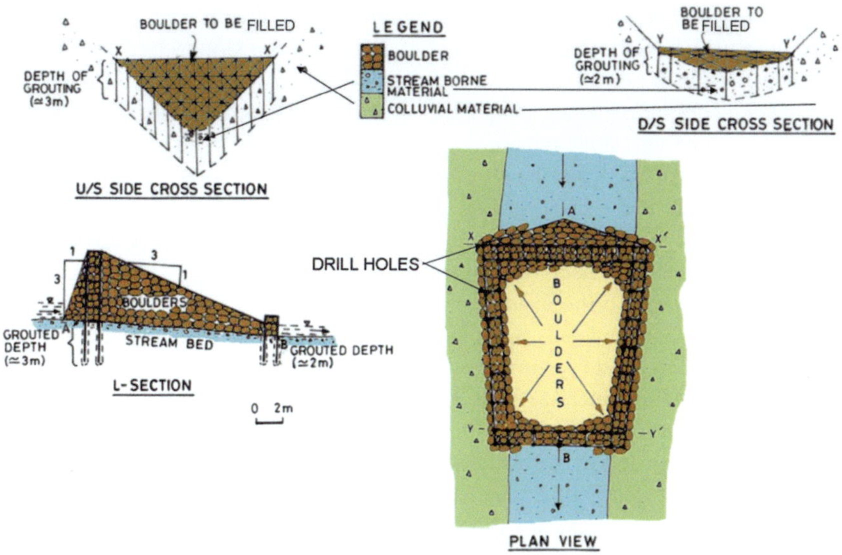

Fig. 11.8 Design of check dams for reducing gradient of Balia stream in Kailakhan

bends where the effect of toe cutting is most prominent (Fig. 11.7). With the passage of time, there will be an accumulation of sediments at the upstream side of these check dams due to development of aggradation condition on the upstream reaches. This will in turn flatten the overall river bed gradient of Balia stream.

11.8 Conclusions

Kailakhan area, situated in the SE corner of Lake Nainital, represents a densely urbanized pocket within Nainital town limit. It is also one of the unstable areas of this hill station, which is facing repetitive slumping problem for many decades in past. Geologically the area is represented by weak Infra-Krol rocks like shales and slates along with phyllites. Rocks exposed here, show well developed foliation planes with one set of steeply dipping joint plane in addition to some random joints. Just below the urbanized stretch of one such slope an extensive slide zone has been identified. The slide has gradually increased in size in the last decades. Presently it is the most dominant slide of this area, threatening the urbanized upper slope and also Nainital–Kathgodam motor road further up on the same slope. Detailed field investigation revealed that this slide provides a classic example where same slope is composed of different types of slope forming material at different levels, for which failure modes are also different. In the upper part which represents a scarp face, toppling failure has been identified as the dominant mode. On the other hand, further down on the loose debris zone, continuing up to stream level, symptoms of slope instability have been clearly observed. In this part of slope, circular failure is identified as the mode of instability. Status of stability of this slope has been calculated by deriving F value using CFC method and SARC program. The stability analysis revealed that slope remains stable in dry condition. With increasing water saturation, particularly under heavy rains, the slope may become unstable. Remedial measures are thought according to the type of slope material and type of instability. For the upper part showing rocky scarp face, grout-anchors can be used to stitch the open near vertical joints. At the same time, it is suggested to reduce slope angle to bring in more stability. For the steep lower slope segment, where debris is accumulated, two different measures of slope stabilization have been suggested. The first type involves protection of valley slopes from toe erosion by providing two tier concrete cubes. The second type involves flattening of stream gradient by constructing boulder fill check dams at suitable stream reaches. The design of the suggested measures has also been indicated. As a general long term treatment for the loose debris zone, biotechnical slope stabilization method using fast growing plant species can also be implemented to restore the natural geo-environmental equilibrium of the slide area. Moreover, this measure will also enhance the aesthetic beauty of presently landslide ravaged condition of Balia valley at Kailakhan. The implementation of these measures will help to tackle future slope stability problems at Kailakhan and thus may act as an effective landslide hazard management strategy.

References

Anbalagan R (1993) Environmental hazards of unplanned urbanization of mountainous terrains: a case study of a Himalayan town. Q J Eng Geol 26:179–184

Anbalagan R, Singh B, Chakraborty D, Kohli A (2007) A filed manual for landslide investigations. Publication of DST (Government of India), New Delhi, p 153

Bishop AW (1955) The use of the slip circle in the stability analysis of earth slopes. Geotechnique 5:7–17

Hoek E, Bray J (1981) Rock slope engineering. Stephen Austin & Sons Limited Publishers, Hertford, p 358

Singh B, Goel RK (2002) Software for engineering control of landslide and tunnelling hazards. A. A. Balkema Publishers, Tokyo, p 344

Valdiya KS (1980) Geology of Kumaun Lesser Himalaya. Wadia Institute of Himalayan Geology, Dehra Dun, p 291

Valdiya KS (1988) Geology and natural environment of Nainital hills. Kumaun Himalaya, Gyanodaya Prakashan, Nainital, p 155

Chapter 12
Natural Hazards of the Arabian Peninsula: Their Causes and Possible Remediation

Arun Kumar

12.1 Introduction

The International Year of Planet Earth (IYPE) was declared for the year 2008, but its activities spread over 2007–2009. It was jointly initiated by the International Union of Geological Sciences (IUGS) and the Earth Science division of the United Nations Education, Scientific and Cultural Organization (UNESCO). Among the ten science themes relevant to society, one of them namely "Hazards – Minimizing risk, Maximizing Awareness" was the focus of extensive worldwide study. Beer (2010) edited a volume of several papers presented at the Hazards Symposium organized at the 33rd International Geological Congress held in Oslo, Norway in 2008, which was part of the IYPE activity. Several papers in this volume provide a comprehensive view on all aspects of natural hazards, their causes, social implications and how developing science and technology is helping in mitigation and future challenges. This paper provides an overview of past and present natural hazards that have impacted the Arabian Peninsula and suggests remedial measures.

The Arabian Peninsula (Latitude 8–32°N; Longitude 36–60°E) is a large area that covers 2.3 million km^2 (Allison et al. 1997) of land in the west Asia, is flanked by the Arabian Gulf in the east, the Gulf of Oman in the southeast, the Red Sea in the west, the Gulf of Aqaba in the northwest, the Gulf of Aden and the Arabian Sea in the south. The Kingdom of Saudi Arabia is the largest country on this peninsula, other countries are, Kuwait, Iraq, Syria and Jordan in the north, Yemen and Oman in the south, and United Arab Emirates (UAE), Qatar and Bahrain in the east (Fig. 12.1). The Arabian Shield is the plateau region in the central and western

A. Kumar (✉)
Center for Petroleum and Minerals, Research Institute, King Fahd University of Petroleum and Minerals, P. O. Box 1945, Dhahran 31261, Saudi Arabia

Department of Earth Science, Ottawa-Carleton Geoscience Centre, Carleton University, 1125 Colonel By Drive, Ottawa, ON K15 5B6, Canada
e-mail: akumar@earthsci.carleton.ca; arunkumarlko@hotmail.com

R. Sinha and R. Ravindra (eds.), *Earth System Processes and Disaster Management*, Society of Earth Scientists Series 1, DOI 10.1007/978-3-642-28845-6_12,

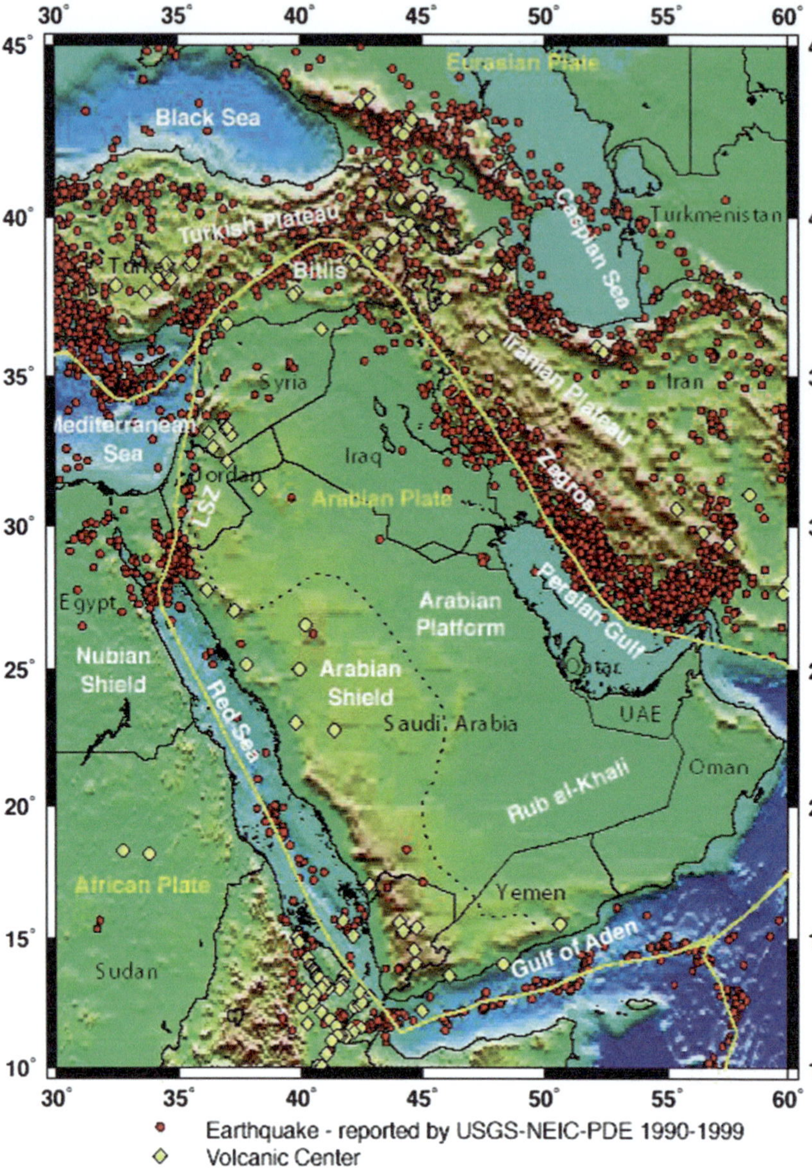

Fig. 12.1 The Arabian Peninsula showing tectonic and geological features. Earthquakes and volcanic centers are shown as red circles and yellow diamond respectively (After Al-Amari et al. 2007. http://www.a-alamri.com/home/files/Full-Projects/earthquaImproving-the-level-of-seiske-location-and-mmagnitude-calibrationic-hazard-parameters-in-Saudi-Arabia-using.pdf)

regions along the Red Sea coast and the plains of the Arabian Platform are in the east. The northern and southern regions are lowlands with several mesas and buttes that are characterized by the Tabuk Basin, the Nafud Basin, the Sirhan-Turayf Basin and the Widyan Basin in the north and great desert of the Rub al-Khali Basin in the south-central regions of the Arabian Peninsula (Grainger 2007).

Oil was discovered in the region during the late 20s (1927 in Iraq) and 30s (1932 in Bahrain and 1938 in Saudi Arabia) that led to the economic boom in this region which continues absolutely unabated even now. Extensive construction work on infrastructure projects and development of new cities are going on and massive new projects are being developed across the region. Under these circumstances, the understanding of possible impact of natural disasters in the region becomes vital for the safety and welfare of the people and property in this region. The author of this paper had earlier cautioned against the potential natural hazards against the reclaimed islands in the Arabian Gulf (Kumar 2009a) and documented historical earthquake and tsunami hazards of this region (Kumar 2009b). Likewise Jordan et al. (2006) and Jordan (2008) provided the history of tsunamis that impacted the Arabian Peninsula. Shehata et al. (1990) discussed three major geological hazards along the Arabian Gulf coast, viz. collapsing, swelling and corrosive behavior of the sabkha environment, erosion, abrasion and deposition problems associated with moving sand and subsidence problems related to sinkholes and offered possible solutions through geotechnical mitigation methods. The present paper chronicles known natural hazards of this region, discusses their origins and suggests possible remedial measures to mitigate, to certain extent, the impact of such hazards.

12.2 Natural Hazards of the Arabian Peninsula

There are several types of natural hazards that have struck this region in the past and continue to strike even now. Natural hazards are broadly geological and atmospheric in origin. Geological hazards include earthquakes, tsunamis and/or tidal waves, volcanic eruption, erosion from currents in the Arabian Gulf and atmospheric hazards include tropical cyclones, sand storms, and sudden cloud burst resulting in flooding. In addition to these hazards, serious problems of ground water depletion, coastal erosion and loss of biodiversity and mangrove forests have already created serious environmental problems for the region (Khan and Kumar 2009). There is a clear possibility that global warming might intensify atmospheric (increased frequency and higher intensities of tropical storms and severe problems of erosion and flooding due to increased events of cloud bursts) hazards in the Arabian region leading to increased threats to the existing infrastructure of the coastal regions and new offshore island communities in the Arabian/Persian Gulf Murty and El-Sabh (1984), Alshahran et al. (2003), Goudie (2003).

12.2.1 Geological Hazards

12.2.1.1 Earthquakes

The Arabian Shield and the Arabian Platform are normally considered aseismic, but western part of the shield areas have experienced seismic activity related to volcanism, rifting and sea-floor spreading in the Red Sea (Ambraseys et al. 2005). Due mainly to opening of northern Red Sea and Gulf of Aqaba, northwestern regions of this peninsula experience higher level of seismic activity (El-Isa and Al-Shanti 1989). The Zagros Fold Belt in Iran is a major source of high magnitude earthquakes that adversely impact eastern coastal regions of the Arabian Peninsula (Ambraseys and Melville 1982; Al-Amri 2005, 2007). According to Grainger (2007) the Arabian Peninsula and the adjoining regions form the Arabian Plate whose tectonic boundaries are formed by several faults. The Baltis Suture and Zagros Suture (strike slip thrust fault) in north and northeast, Oman Ophiolite complex in southeast, Dead Sea Transform Fault in northwest, Red Sea spreading center (a divergent margin) in west, and Gulf of Aden spreading center or transform faults of Gulf of Aden, East Sheba Ridge and Owen Fracture Zone (strike slip fault) in south make the tectonic boundary of the Arabian Plate (Fig. 12.2).

In western region of the Arabian Peninsula seismic activity is associated with Quaternary volcanism and divergent plate boundary of the Red Sea. Some of the recent earthquakes in the region are mentioned here. According to the United States Geological Survey online report (http://earthquake.usgs.gov/earthquakes/world/middle_east/seismicity.php) six earthquakes of 3–4.5 M hit the Harrah Al-Shaqah area of Al-Ais region around Al-Medina in western Saudi Arabia on May 17, 2009 and on May 19, 2009 the largest 5.7 M earthquake hit the region followed by several aftershocks. The Civil Defense department of the Saudi Arabian Government warned people living in the radius of 40 km of the epicenter to leave their homes and according to "Arab News" dated May 19, 2009, 2,289 persons were evacuated from villages in Al-Ais area. This event was widely reported in media of the Arab countries. In September, 2005 an earthquake measuring 3.7 on Richter scale hit the holy city of Mecca and neighboring Otaibah. The town of Haradh in Eastern Province of Saudi Arabia was hit by an earthquake of 4.1 on Richter scale and an earthquake of similar intensity had hit this area in 2005 also (see online report on "Earthquake Hits Saudi Arabia" dated November 23, 2006 by Ali Mateer; http://www.aawsat.com/english/news.asp?section=1&id=7133).

Kumar (2009a, b) reviewed information on earthquake events in the Arabian Gulf and discussed potential earthquake hazard to the several new offshore townships on reclaimed islands in the Arabian Gulf. Major urban centers in the Arabian Gulf coastal regions like Dubai and other cities are prone to dangers of seismic activity despite the fact that Arabian Peninsula is considered seismically to be a relatively stable region. According to Al-Amri (2005, 2007) there are 25 seismogenic source zones in the Arabian Peninsula and adjacent countries. They are rift systems, strike slip and normal faults, joints, lineaments, subduction and

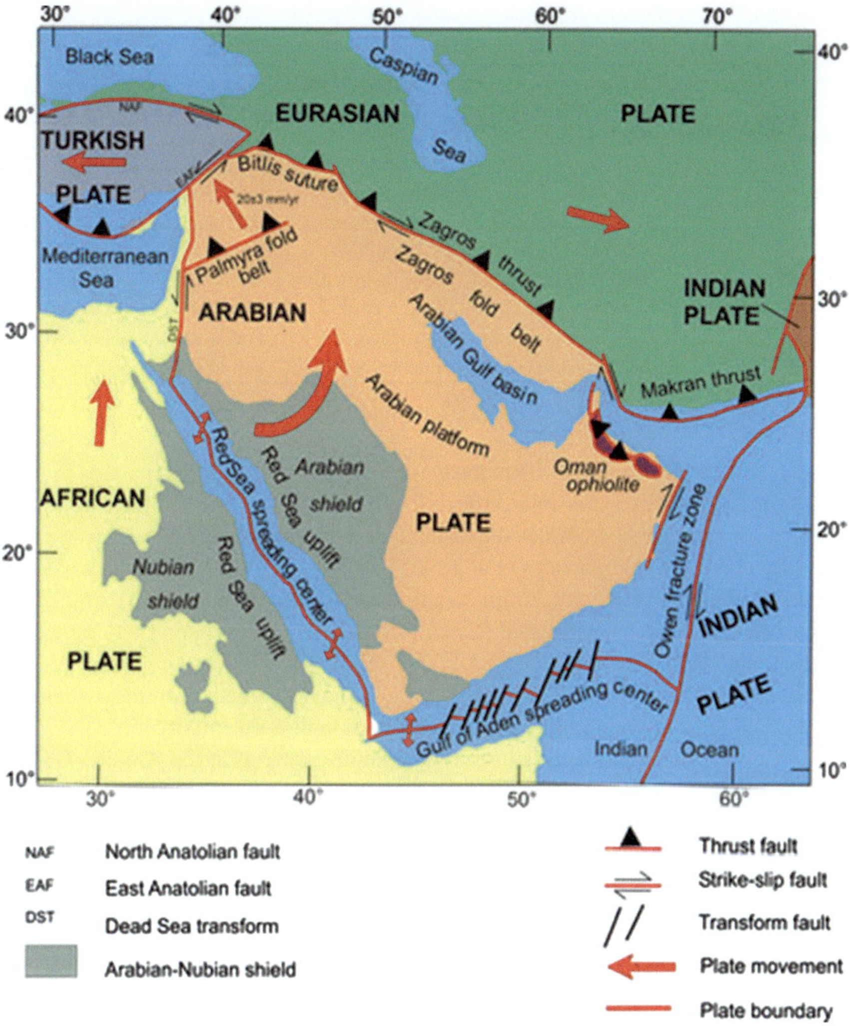

Fig. 12.2 Plate-tectonic relationships of the Arabian Plate (After Grainger 2007) (http://feww.wordpress.com/2009/11/05/arabia-shakes)

collision zones (Figs. 12.1 and 12.2). The Zagros Thrust Fault demarcates the plate boundary between the Arabian and the Eurasian tectonic plates and is a seismically active zone (Fournier et al. 2006). Large ($M \leq 5$) earthquakes occur frequently along this thrust (Fig. 12.1). Even low frequency ground motion amplification could be significant because it can excite vibrations in tall buildings (Rodgers 2005). On September 11, 2008 a 6.2 M earthquake jolted the Iranian port city of Bandar Abbas. Its epicenter was in the Arabian Gulf 43 km west of Qeshm Island and a tsunami warning was issued. This earthquake shook buildings all over the UAE and high rise towers were specially affected. Earthquakes originating from Iranian

Table 12.1 List of tsunamis generated by earthquakes within the Arabian Gulf region

Date	Location	Tsunami	Comments
June 17, 978	Siraf, Iran	Tsunami, coastal flooding	Over 2,000 people killed, sank ships
Spring, 1008	Siraf, Iran	? tsunami or a storm surge	Large waves sank ships, people killed
Nov., 1426	Near Bahrain	None recorded	Earthquake
Sept. 1871	? earthquake at Bushehr, Iran	Tsunami, coastal flooding	Al-Tabah event, ships sunk, people killed
~1832	Hufuf, Saudi Arabia	None recorded	Earthquake
June 13, 1858	Bushehr, Iran	None recorded	Earthquake
May 19, 1884	Qeshm Island, Iran	None recorded	Felt at Ra's al-Khaima
Nov. 27, 2005	Qeshm Island, Iran	None recorded	Felt in Dubai, Sharjah

Coastal Source Zone (100–400 km away) could cause catastrophic damage to high rise buildings of UAE. Seismic hazard of tall non earthquake resistant buildings pose high level of risk from distant earthquakes (Petrovski 2005). Ambraseys and Melville (1982) and Ambraseys et al. (2005) provide an excellent historical record of earthquakes in Iran, Egypt, Arabian region and the Red Sea. Several earthquakes originating in Iran have generated tsunamis that have severely impacted the whole Arabian Gulf region (Table 12.1).

The danger of earthquakes would be significantly higher for the reclaimed islands and their tall buildings because reclaimed land is potentially more susceptible to liquefaction and slope failure. Reclamation material in the Arabian Gulf is mainly dredged carbonate sand and little is known about underlying or nearby fracture zones and seismicity (Williams 2005). All these factors make tall buildings constructed on reclaimed islands quite dangerous and prone to seismic destruction.

12.2.1.2 Tsunamis and Tidal Waves

Jordan et al. (2006) and Jordan (2008) documented historical tsunami events of the Arabian Peninsula including one doubtful tsunami in the Arabian Gulf (Spring A.D. 1008, Siraf, Iran) and following two localized events in the Red Sea. (1) A.D. 1879, 11 July, Tor (present day El Tor), Egypt, Moderate earthquakes were felt in upper Egypt, village of El Tor was inundated by tsunami, and (2) A.D. 1884, 20 July, Massawa (present day Mitsiwa), Eritrea, offshore earthquake occurred, sea waves built up in harbor and swept over causeway.

Kumar (2009a, b) documented possible tsunami events in the Arabian Gulf and discussed their origins and impacts in the region. Since the Arabian Gulf is an inland sea with an opening to the Arabian Sea through the narrow Strait of Hormuz, it is to an extant protected from tidal waves, storm surges and tsunamis generated in the Arabian Sea or the Indian Ocean. However some tsunami energy would always leak into the Arabian Gulf region (Murty et al. 2007; Kumar and Alam 2010). Bhaskaran et al. (2007) calculated tsunami travel time for 250 locations in 35

Table 12.2 List of tsunamis generated by earthquakes of the Indian subcontinent that potentially had impacted the Arabian Gulf region

Date	Location	Cause
326 B.C.	Kachchh region/Indus Delta	Earthquake
April 1–May 9, 1008	Iranian coast in Arabian Gulf	Earthquake
1524 A.D.	Off Dabhol coast, Maharashtra, India	Earthquake
June 16, 1819	Kachchh region, Gujarat, W. India	Earthquake
June 19, 1845	Kachchh region, Gujarat, W. India	Earthquake
Nov. 28, 1945	Makran Coast, Southern Pakistan	Earthquake

countries around the Indian Ocean for December 26, 2004 Sumatra Earthquake. The authors included Manama (Bahrain), Bandar Abbas, Bander-e-Bushehr and Jask (Iran), Kuwait City (Kuwait), Doha and Dukhan (Qatar), Abu Dhabi, Dubai, and Sharjah (UAE) from the Arabian Gulf region. This demonstrates that tsunamis generated in the Indian Ocean do travel in the Arabian Gulf. The Kachchh region of Gujarat in western India and the Makran coast of southern Iran and Pakistan have experienced tsunamigenic earthquakes in the past and most likely such events had impacted coastal regions of the Arabian Gulf also, however no such records were found.

Rastogi (2007) and Rastogi and Jaiswal (2006) compiled a comprehensive list of tsunamis of the Indian Ocean region and listed tsunamigenic earthquakes in the Arabian Sea that potentially had impacted the coastal regions of the Arabian Gulf (Table 12.2). Similarly Bou-Rabee and VanMarcke (2001), Jordan et al. (2006) and Jordan (2008) compiled lists of earthquakes and possibly related tsunami events in the Arabian Gulf (Table 12.2). Bou-Rabee and VanMarcke (2001) specially mention the September, 1871 "Al-Tabah" tsunami event in which most Kuwaiti trading ships (and most likely ships belonging to other countries also) were sunk in the Arabian Gulf and the Arabian Sea causing massive coastal flooding all along the Arabian Gulf leading to high human mortality. Most probably "Al-Tabah" tsunami event was caused by an earthquake at Bushehr, Iran on the same date.

The tsunamigenic Makran earthquake of November 28, 1945 was deadliest on record for the South Asian region. It generated wave heights up to 11.5 m (Dimri and Srivastava 2007) and run-ups up to 17 m on the Makran coast (Murty et al. 1999; Pararas-Carayan 2006). More than 4,000 people were killed along the Makran coast of southern Iran and Pakistan and few in Kachchh region of Gujarat in western India and in Mumbai. This tsunami affected the Iranian coast in the Arabian Gulf region also (Chadha 2007; Pararas-Carayan 2006). Makran tsunami waves also swept into the Arabian Gulf and washed out a large sandbar at Ras al-Khaimah, UAE (Jordan 2008). I am not aware of any written record of impact of 1945 Makran coast tsunami in the Arabian Gulf coastal regions, but according to anecdotes from several fishing villages between Dammam and Jubail along the east coast of Saudi Arabia, several fishermen died due in a series of wave upsurges on the Arabian Gulf coastal regions of Saudi Arabia. According to Pararas-Carayan (2006), *"seismotectonics of the Makran subduction zone, historical earthquakes in the region, the recent earthquake of October 8, 2005 in Northern Pakistan, and the great tsunamigenic earthquakes of December 26, 2004 and March 28, 2005, are indicative of the active tectonic collision process that is taking place along the entire southern and southeastern boundary of the Eurasian plate as it collides with the Indian plate and adjacent microplates thus the Makran subduction zone has a relatively high potential for large tsunamigenic earthquakes"*. The earthquakes occur quite frequently in Iran and adversely impact the Arabian Gulf region. But most such earthquakes are not tsunamigenic. However, future occurrence of tsunamigenic earthquakes along the Iranian coast of the Arabian Gulf cannot be ruled out.

Mokhtari et al. (2008) used 2D seismic reflection data to delineate the regional tectonic elements of the Makran region and the Gulf of Oman and concluded that this region is prone to future tsunamis. A review of archival research has shown that average repeat period of magnitude 8+ earthquakes along the Makran Subduction Zone (MSZ) is between 100 and 250 years and this region is capable of producing both tectonic and volcanic tsunami (Heidarzadeh et al. 2008). The numerical modeling of 1945 Makran tsunami indicates that in case of a similar tsunami event Iran and Pakistan will experience wave heights of up to 5–7 m, 5 m in northern coast of Oman and 2 m in the eastern coast of the United Arab Emirates (Heidarzadeh et al. 2009). In case such events occur they will seriously impact reclaimed island communities being developed in the Arabian Gulf.

12.2.1.3 Volcanism

Volcanic eruptions are both awesome and deadly events, and are major natural hazard. In addition, submarine volcanism has been responsible for past several tsunamis. Pyroclastic eruption from a volcano causes serious damages in the surrounding regions by falling hot ash, dust, and smoke within minutes to hours, molten lava spewed from a volcano can ignite fires in nearby forests and human settlements. Volcanic ash mixed with heavy rains and melting glaciers produce mud that can flow for miles, overrunning infrastructure facilities and human settlements large or small. Large plumes of ash and gas (mainly sulphur dioxide and carbon dioxide) ejected high into the atmosphere can influence climate, sometimes on a global scales. A recent example is a series of eruptions of the Eyjafjallajöekull volcano, Iceland in March/April, 2010 which caused worldwide disruption to air travel especially across western and northern Europe. This led to enormous difficulties to travelers and huge financial losses to airline companies.

The volcanoes of the Arabian Peninsula are mostly located in the western regions and on the southern end of the Red Sea off Yemeni coast (Fig. 12.1). There are widespread basaltic lava flows that are popularly known as "harrat" in Saudi Arabia (Figs. 12.3 and 12.4). These harrats cover approximately 180,000 km^2 from Yemen in south to Turkey in North. The Global Volcanism Program initiated by Siebert and Simkin (2002) lists 18 volcanoes on the Arabian Peninsula and 4 on the southern end of the Red Sea providing their location, summit elevation and last known eruption (Tables 12.3 and 12.4). Most of these are volcanic fields and the last

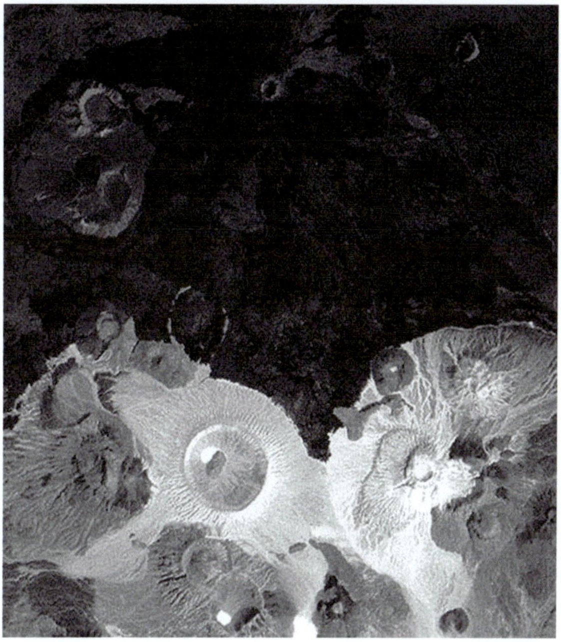

Fig. 12.3 Black and white volcanoes of Saudi Arabia and volcanic fields (Harrat) (After Siebert and Simkin 2002: http://www.volcano.si.edu/world/)

Fig. 12.4 Jebel Bayda volcano of Saudi Arabia (After Siebert and Simkin 2002: http://www.volcano.si.edu/world/)

known eruption on the Arabian Peninsula was in 1937. During the past 4.5 ka there have been 13 major eruptions, averaging one every 346 years (Siebert and Simkin 2002).

Volcanoes are of various shapes and sizes, and their morphological characterization has been documented by various workers (Simkin and Siebert 1994; Siebert and Simkin 2002). These workers define the status of volcanoes as "Historical" because eruption was documented during or shortly after observation, or

Table 12.3 Volcanoes of the Arabian Peninsula (Compiled from http://www.saudiaramcoworld.com/issue/200602/volcanic.arabia.htm)

Volcano name	Volcano type	Volcano status	Latitude	Longitude	Location	Last known eruption	Summit elevation
Al Harrah	Volcanic field	Holocene	31.08°N–31°5′0″N	38.42°E–38°25′0″E	Saudi Arabia	Unknown	1,100 m
HarratarRahah	Volcanic field	Anthropology	27.80°N–27°48′0″N	36.17°E–36°10′0″E	Saudi Arabia	Unknown	1,950 m
Harrat 'Uwayrid	Volcanic field	Anthropology	27.08°N–27°5′0″N	37.25°E–37°15′0″E	Saudi Arabia	640 A.D. (?)	1,920 m
HarratLunayyir	Volcanic field	Historical	25.17°N–25°10′0″N	37.75°E–37°45′0″E	Saudi Arabia	1000 (in or before)	1,370 m
HarratIthnayin	Volcanic field	Holocene	26.58°N–26°35′0″N	40.20°E–40°12′0″E	Saudi Arabia	Unknown	1,625 m
HarratKhaybar	Volcanic field	Historical	25.00°N–25°0′0″N	39.92°E–39°55′0″E	Saudi Arabia	650 A.D. ± 50 years	2,093 m
HarratRahat	Volcanic field	Historical	23.08°N–23°5′0″N	39.78°E–39°47′0″E	Saudi Arabia	1256	1,744 m
HarratKishb	Volcanic field	Holocene	22.80°N–22°48′0″N	41.38°E–41°23′0″E	Saudi Arabia	Unknown	1,475 m
Harrat al Birk	Volcanic field	Holocene	18.37°N–18°22′0″N	41.63°E–41°38′0″E	Saudi Arabia	Unknown	381 m
JabalYar	Volcanic field	Historical	17.05°N–17°3′0″N	42.83°E–42°50′0″E	Saudi Arabia	1810 ± 10 years	305 m
Harra of Arahab	Volcanic field	Historical	15.63°N–15°38′0″N	44.08°E–44°5′0″E	Yemen	500 A.D. ±100 years	3,100 m
Jabal El-Marha	Tuff cone	Holocene ?	15.245°N–15°14′42″N	44.236°E–44°14′11″E	Yemen	Unknown	2,506 m
JabalHaylan	Volcanic field	Anthropology	15.43°N–15°26′0″N	44.78°E–44°47′0″E	Yemen	1200 B.C. (in or after)	1,550 m
Harras of Dhamar	Volcanic field	Historical	14.57°N–14°34′0′N	44.67°E–44°40′0′E	Yemen	1937	3,500 m
Unnamed	Submarine	Uncertain	12.25°N–12°15′0″N	45.00°E–45°0′0″E	Gulf of Aden	Unknown	Unknown
Harraes-Sawad	Volcanic field	Historical	13.58°N–13°35′0″N	46.12°E–46°7′0″E	Yemen	1253	1,737 m
Harra of BalHaf	Volcanic field	Holocene	14.05°N–14°3′0″N	48.33°E–48°20′0″E	Yemen	Unknown	233 m
BirBorhut	Volcanic field	Holocene ?	15.55°N–15°33′0″N	50.63°E–50°38′0″E	Yemen	Unknown	Unknown

Table 12.4 Volcanoes of the Red Sea (Compiled from http://www.volcanolive.com/redsea.html)

Volcano name	Volcano type	Latitude	Longitude	Location	Known eruptions	Summit elevation
Jebel al-Tair	Stratovolcano	15.70 N	41.742 E	Red Sea, Yemen	2007–2008, 1883, 1863, 1833 ± 1, 1750 ± 50	244 m
Jebel Zubair	Shield volcano	15.05 N	42.18 E	Red Sea, Yemen	1846?, 1824	191 m
Zukur	Shield volcano	14.02 N	42.75 E	Red Sea, Yemen	Holocene	624 m
Hanish	Shield volcano	13.72 N	42.73 E	Red Sea, Yemen	No recent eruptions	422 m

"Anthropology" for volcanoes that are recent but undated and described in either legends or dated by buried artifacts and "Holocene" for volcanoes whose eruption is certain but undated.

Well known volcanic eruptions of the Arabian Peninsula are as follows; 1937 near Dhamar in northern Yemen, 1846 in Red Sea Island Saddle in the Zubair Islands 90 km northwest of Yemeni port city of Hodaida. Harrigan (2006) provides a graphic account of "The Madinah Eruption" of 1256 that lasted for 52 days and how it destroyed the surroundings of the holy city of al Madinah in Saudi Arabia. There are evidences of eruption and lava flows dating ~4.5 ka in the region.

Seach (2010) provides most complete and current information about the Red Sea volcanoes (Table 12.4). The most recent volcanic eruption occurred on Jebel al-Tair, a volcanic island 150 km away from the port city of Hodeida, Yemen, on 30th September 2007. Lava flowed into the sea and this event lasted for over 3 days. Several earthquakes occurred before, and one measuring 7.3 on the Richter scale was recorded a day before the eruption. According to Seach (2010) "ash and pumice rafts floated 10 km from the island. On 3rd November 2007, fire fountains were visible on the NW side of the island near the summit. Satellite images showed hotspots at the volcano until June 2008." This eruption produced 1 km long lava flow and blackened the water with in a 9.7 km radius of the island. Eight people including three personnel from the Yemeni army perished in this event.

12.2.2 *Atmospheric Hazards*

12.2.2.1 Tropical Cyclones and Storm Surges

Murty (1999) describes the relationship of storm surges to tropical cyclones as, *Storm surges are oscillations of the water level in the coastal water body due to the tangential wind stresses and atmospheric pressure gradients mainly associated with the meteorological forcing fields of traveling synoptic scale weather systems such as tropical cyclones (TCs) and Extra-Tropical Cyclones (ETCs).* In the higher latitudes of northern hemisphere storm surges are generated by ETCs which generally travel from west to east with some north to south component. Similarly, surges are generated in the lower latitudes by TCs that travel mostly east to west with some south to north component. In a peak storm surge the tangential wind stress pushes the water towards the coast up to 100 km leading to sudden inundation and flooding of coastal regions by the sea water (Dube et al. 1997; Murty 1999). The Arabian Sea is normally under the influence of tropical cyclones and storm-surges and cyclones occur either during pre-monsoon or post-monsoon season but rarely during the monsoon season (Murty and El-Sabh 1984). Tropical cyclones that may impact the Arabian Peninsula are generated in the eastern part of the Arabian Sea and sometimes they can from in the Bay of Bengal also and move across the Indian Peninsula and get re-intensified over the Arabian Sea on their way to the peninsula.

According to El Sabh and Murty (1989) the Arabian Gulf and the Red Sea are mostly influenced by extra-tropical weather systems. The west to east directed extra-tropical cyclones and the generally east to west directed tropical cyclones converge near the Strait of Hormuz where a meso-scale weather system "winter Shamal" develops and the Arabian Gulf is subject to major storm surges. Strong winds associated with the Shamal system, coupled with atmospheric pressure gradients, topography and tidal effects, can give rise to water level deviations of several meters. Storm surges observed during January 17–19, 1973 were widespread in the Arabian Gulf (El-Sabh and Murty 1989).

It is generally believed that the countries of Arabian Peninsula appear to be "safe" from the devastation of tropical storms generated in the Arabian Sea. However, the tropical cyclone Gonu (Fig. 12.5) proved that such severe events do occur in the Arabian Sea and can also be devastating. On June 5, 2007 the intensity of cyclone Gonu had reached a Category 5, with sustained winds measuring 160 miles per hour (http://www.nasa.gov/vision/earth/lookingatearth/gonu.html). According to the Joint Typhoon Warning Center (http://earthobservatory.nasa.gov/NaturalHazards/natural_hazards_v2.php3?img_id = 14328) it was the most powerful cyclone ever to threaten the Arabian Peninsula since record keeping began in 1945. This cyclone had weakened by the time it hit Oman's capital city Muscat. It still caused enormous damage to the city infrastructure and loss of life and property. Another tropical cyclone 03B followed Gonu. It originated in the Bay of Bengal on June 25, 2007 and crossed the Indian Peninsula and reformed in the Arabian Sea south of Pakistan with wind speed at 40 miles per hour. It caused flooding and damage to Karachi, Pakistan where around 200 people died. Storm surge from cyclone 03B (also named Yemyin) was predicted to be moderately high, even though the storm was not strong, because the offshore waters are shallow. In 2006, tropical storm Mukda was the only tropical system to form in the Arabian Sea and it remained well out in the sea before dissipating.

In 2010, three cyclonic storms were generated in North Indian Ocean Cyclone Season, first cyclone Laila was generated in the Bay of Bengal and had a land fall along the Andhra Pradesh on the east coast of India. This caused heavy loss of life and property. Second cyclone Bandu was active during May 19–23 in the Arabian Sea with 50 mph wind velocity. According to Wikipedia (http://en.wikipedia.org/wiki/Cyclone_Phet) this cyclone impacted the coast of Somalia and the Yemeni

Fig. 12.5 Cyclone Gonu, June 5, 2007, over southeast Oman and northern Indian Ocean (http://www.nasa.gov/vision/earth/lookingatearth/gonu.html)

island "Abed Alkorey" in the Gulf of Aden. It caused death of a girl and destruction of many houses in the island. On May 20 the storm disabled a cargo vessel, MV Dubai Moon, and left it drifting off the Somali coast where 23 crew members were rescued by helicopters from the Royal Navy frigate HMS Chatham. The cargo ship later sank. The third tropical cyclone named Phet (Fig. 12.6) was active during May 31 to June 6, was very severe cyclonic storm of category 4 at its peak intensity with 90 mph wind impacted Oman on June 4 and Pakistan on June 6. It left 44 people dead and loss of property worth 780 million dollars.

Murty and El-Sabh (1984) consider that tropical cyclones generated in the Indian Ocean rarely impact the Arabian Gulf or the Arabian Peninsula, they state (p. 667), *The important point made here is that only rarely, if at all the tropical cyclones from the Arabian Sea travel towards the Gulf of Oman and the Arabian Gulf.* Although recent observations on the movement of tropical cyclones Gonu (Fig. 12.5; June 5, 2006) and Phet (Fig. 12.6; June 5, 2010) and few others observed during the past few years do not support this view. This may be due to impact of global warming or a certain cyclic meteorological phenomenon when the frequency Indian Ocean tropical cyclones increasingly move northward impacting the Arabian Gulf and the Arabian Peninsula. In case such tropical storms continue to impact the Arabian region, there is every possibility that future cyclones may impact offshore island communities developed in the Arabian Gulf especially several Palm Islands off Dubai. This is evident from the fact that Cyclone Phet moved further north than Cyclone Gonu.

Fig. 12.6 Cyclone Phet, June 5, 2010 over eastern Oman and Gulf of Oman (http://www.nasa.gov/vision/earth/lookingatearth/phet.html)

12.2.2.2 Sand and Dust Storms

Occurrences of sand and dust storms are common over the Arabian Peninsula. Desert dust usually consists of small particles (<0.0525 mm) deflated from desert sediments that can be carried to <100 to >10, 000 km (Middleton 1997). Dust particle size varies considerably, smaller the particle size, larger the distance it gets carried away and longer periods (sometimes over a week) it remains suspended in the troposphere. Quartz is the main constituent of the dust, other minerals found are feldspar, calcite, dolomite, micas, clay minerals, gypsum, halite, opal, amorphous inorganic and organic material (Pye 1987). The dust particles are lifted from ground and the factors that control the character of storms are nature of the wind, nature of the soil or sediment, and the presence of any surface obstacles to wind flow (Middleton 1997). Dust storms in northern Africa, the Middle East and Asia put >200–5,000 million metric tons of mineral dust into the Earth's atmosphere every year (Tegen and Fung 1994). An operational model to forecast dust storms was developed for the United States Air Force Weather Agency with probability of detection of dust storm occurrence exceeding 68% over North Africa and 61% for southwest Asia (Barnum et al. 2004).

Haboob is an Arabic word term for intense sandstorms observed in the arid regions of the world including the Arabian region. Such winds in the Arabian Peninsula are frequently created by the collapse of thunderstorms and when they reach the ground, dry loose sand from the desert creates a wall of sediment preceding the storm cloud as exemplified by the March 10, 2009 storm over Riyadh, Saudi Arabia (Fig. 12.7). Middleton (1986) and Washington et al. (2003) provide

Fig. 12.7 Dust storm over Riyadh, March 10, 2009 (http://www.dailymail.co.uk/news/worldnews/article-1160962/Pictured-The-moment-awe-inspiring-desert-storm-engulfed-Saudi-capital.html)

excellent reviews on dust storms in the Middle East and their range of environmental impacts. Significant detailed information about the sand and dust storms specially for the Arabian Peninsula can be found in McKee (1979), Thomas (1997), Alsharhan et al. (1998, 2003).

There are two wind systems that interact over the Arabian Peninsula. The northwesterly winter Shamal trade winds (Fig. 12.8) and the southwesterly summer winds associated with monsoon circulation from the Indian Ocean (Glenie and Singhvi 2002). Shamal wind blows down the Arabian Gulf towards UAE and then turns clockwise across the Rub Al-Khali desert (Fig. 12.9). These winds also create waves and surface currents and transport terrigenous sediments including pre-Holocene dolomitic dust to the marine environments. According to Ali (1994) the summer Shamal is a strong cold front that passes over mountains of Turkey and picks up dust and sand along the way mostly from Jordan and Syria. Its temperatures at the lower elevations remain at ~42°C. The winter Shamal is

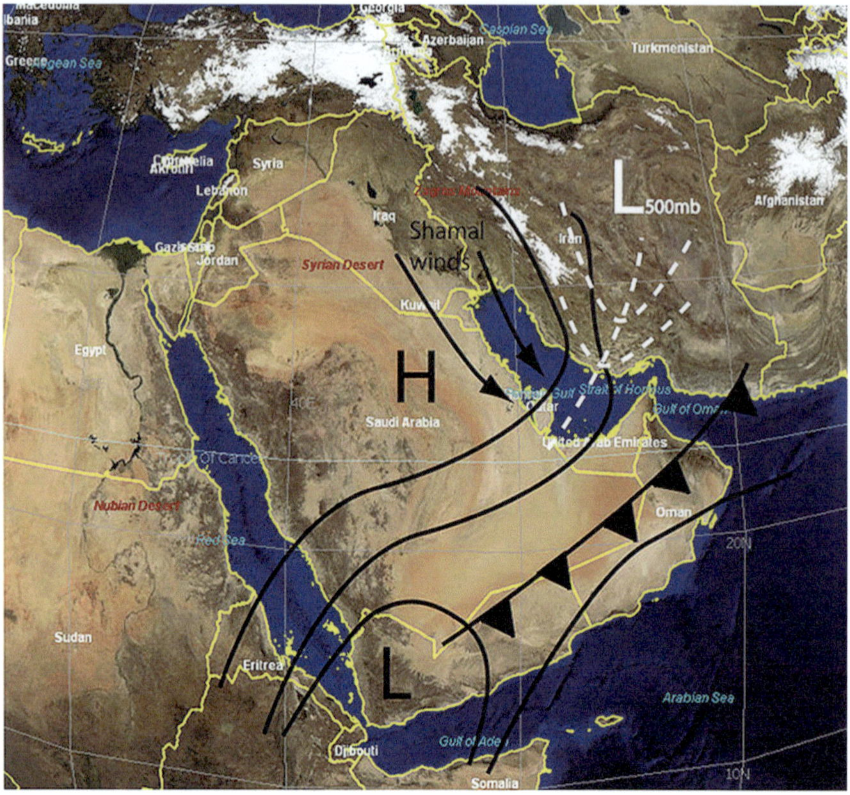

Fig. 12.8 Synoptic conditions leading to the 3–5 days Shamal wind (http://en.wikipedia.org/wiki/Shamal)

associated with the strengthening of a high pressure over the Arabian Peninsula after the passage of a cold front while a deep trough of low pressure maintains itself over areas east of the Arabian Gulf. This leads to strong northerly wind over the Gulf for periods up to 5 days. These winds are associated with cold temperatures. Shamal winds usually last for 3–5 days and may occur anywhere from one to several times per year, mostly in summer but sometimes in winter also.

Dust storms are several thousand feet thick and adversely impact airline services, airports shut down, disruption of road transport, shipping and fishing in the Arabian Gulf and the Red Sea gets affected and sand blasting has also been reported to strip the paint off of cars. Dust storms are known to cause respiratory and eye problems and spread diseases across the globe by blowing virus spores into the atmosphere and spreading them all around. They also cause soil loss from the dry lands and they preferentially remove organic matter and the nutrient-rich

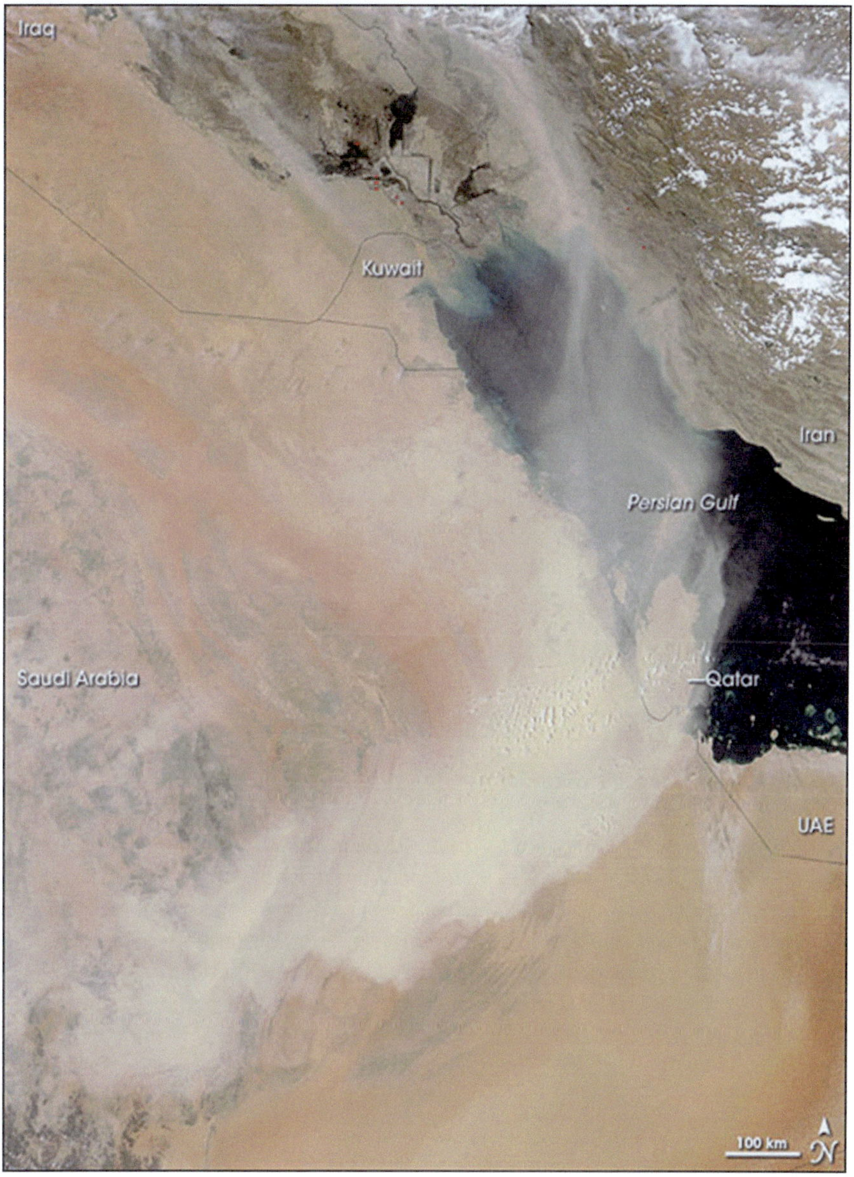

Fig. 12.9 Dust storm over the Arabian (Persian) Gulf and the Arabian Peninsula on September 12, 2008 (http://www.eosnap.com/?s=Dust+storm+Arabian+peninsula)

lightest particles, thereby reducing agricultural productivity. They can also have beneficial effects on agriculture by distributing mineral nutrients. The following are some of the recent examples of sand and dust storms in the Arabian region.

1. October 22, 2004: Northern Saudi Arabia
2. April 12, 2007: Arabian Gulf
3. February 19, 2008: Kuwait
4. May 20, 2008: United Arab Emirates
5. June 6, 2008: Iraq, Kuwait and Saudi Arabia
6. February 12, 2009: Entire Arabian Peninsula
7. March 10–11, 2009: Worst ever in Saudi Arabia
8. May 6, 2010: Gulf of Aqaba and Red Sea
9. June 10, 2010: Red Sea

12.2.3 Other Hazards

Under this heading I would briefly discuss those natural hazards in which natural causes coupled with human activities are involved in creating hazardous conditions that are adversely impacting on human society, fragile ecosystem, and habitats of flora and fauna of the region.

12.2.3.1 Droughts and Desertification

Drought is by far the most damaging of all natural disasters. Worldwide, since 1967, drought is responsible for millions of deaths and has cost hundreds of billions of dollars in damage. Desertification is land degradation in arid environments caused primarily by climatic variations and human activities (Thomas 2003). The issue of global warming induced by wide varieties of anthropogenic activities may be controversial, but average global temperatures have been undisputedly rising for the past few decades. This leads to reduction in availability of soil moisture due to reduced precipitation and increased evapotranspiration resulting in severe drought conditions in an already arid Arabian Peninsula. Goudie (2003) and Squires (2003) provide in-depth analysis of impact of global warming on desertification. Among the human activities that are causing severe environmental stress in this region are unsustainable irrigation practices, overgrazing, uncontrolled cultivation, wood-gathering for fuel, soil salinization and water logging (Alsharhan et al. 2003). Verstraete et al. (2008) edited a volume of the journal "Global and Planetary Change" on climate change and desertification. This volume has several papers dealing with a wide range of ideas that cover the natural and anthropogenic causes of desertification and remediation issues of the arid regions of the world. In arid regions like the Arabian Peninsula biodiversity is very low but the region has precious limited coastal mangrove forests. It should make a lot of sense to local planners to protect this rich habitat for its flora and fauna. Unfortunately, these coastal mangrove forests are being systematically destroyed both along the Arabian Gulf coast (Khan and Kumar 2009) and the Red Sea coast (Kumar et al. 2010; Khan et al. 2010) for urban and infrastructure development.

Fig. 12.10 Aftermath of November 25, 2009 flash floods in Jeddah, Saudi Arabia (http://www.dailymail.co.uk/news/worldnews/article-1233836/Desert-kingdom-Saudi-Arabia-begins-clear-FLOODS-kill-116.html)

12.2.3.2 Flash Floods

Flash floods are sudden overflow of water over land caused by excessive rainfall in a short period of time ranging from few minutes to few hours. They are raging torrents of water after heavy rains that break through river banks, city streets sweeping everything from trees, buildings and bridges whatever comes along (Fig. 12.10). They are unpredictable and may occur within minutes or a few hours of excessive rainfall. Flash floods are unpredictable and appear as a wall of water, which can be 10–20 ft high. Heavy rainfall from a stationary front, hurricane, tropical storm or thunderstorm can cause flash floods. Average annual rainfall in coastal Arabia is <5 cm thus fluvial influx is very low and is limited to occasional flooding of desert wadis due to rains. Table 12.5 shows few of the more recent examples of flash flooding events in the Arabian Peninsula.

12.3 Discussion

Spiess (2008) in his article discussing the role of government policy on climate change and sustainable development in the Arabian countries sates, *Intergovernmental Panel on Climate Change (IPCC) has emphasized that understanding the institutional context in which policies are made and implemented is critical to define sustainable development paths from a climate change perspective. Nevertheless, while the importance of social, political and cultural factors is getting more recognition in some parts of the world, little is known about the human dimensions*

Table 12.5 Recent flash flood events on the Arabian Peninsula

Date	Location	Death/ injuries	Property damage	Source of information
April 29, 2005	Jeddah, Saudi Arabia	30 killed	Large numbers of houses and automobiles	http://news.bbc.co.uk/2/hi/middle_east/4497443.stm
November 25, 2009	Jeddah, Saudi Arabia	116 people including 77 Hajj pilgrims	Damaged 8,092 homes and crushed 7,143 vehicles	http://www.dailymail.co.uk/news/worldnews/article-1233836/Desert-kingdom-Saudi-Arabia-begins-clear-FLOODS-kill-116.html
December 12 and 13, 2009	Abu Dhabi and Dubai, UAE	6 killed	Large numbers of houses and automobiles	http://www.moveoneinc.com/blog/relocations/excessive-rainfall-hits-uae
April 17, 2010	Southwestern Saudi Arabia	250 people rescued, no death report	No significant damage	http://2012info.ca/EarthWatch/?p=93369
May 3, 2010	Southwestern Saudi Arabia	9 killed	275 car accidents in Riyadh	http://english.aljazeera.net/news/middleeast/2010/05/20105410122731245.html
July 16, 2010	Southwestern Saudi Arabia and Yemen	23 killed in Yemen and 1 in Saudi Arabia	Large numbers of houses, one hospital and automobiles	http://www.worldbulletin.net/news_detail.php?id=61411

or the contexts in which they operate in the affluent oil economies of the Arabian Peninsula. Policies that implicitly subsidize or support a wasteful and environmentally destructive use of resources are still pervasive, while noteworthy environmental improvements still face formidable political and institutional constraints to the adaptation of the necessary far reaching and multisectoral approach. This statement is very true if one looks at the fast pace of construction of roads, ports, and other infrastructure facilities, new cities, high rise towers, reclaimed offshore island communities, recreations facilities etc. Environmental concerns like destruction of coastal mangrove forests (Khan and Kumar 2009), coral reefs and wide varieties of coastal environments have normally been disregarded in the construction process (Kumar 2009a). As an example of poor environmental record of the countries of the Arabian Peninsula, the data from the 2005 Environmental Sustainability Index (ESI) that benchmarks the ability of 146 nations to protect their environment over the next several decades by comparing five fundamental components of sustainability: Environmental Systems; Environmental Stresses; Human Vulnerability to Environmental Stresses; Societal Capacity to Respond to Environmental Challenges; and Global Stewardship. While there was insufficient data for Qatar and Bahrain, Kuwait ranked 138, Saudi Arabia 136 und the UAE 110 (Yale Center for Environmental Law and Policy/Center for International Earth Science Information Network, 2005 in Spiess 2008).

As described above, natural disasters have impacted the Arabian Peninsula during the past and they are impacting adversely even now and there is enough scientific data to support the idea that such natural hazards will impact this region in future also. It is quite clear that there is an urgent need to frame joint disaster management and prevention policies by all governments of this region to minimize impacts of various types of natural disasters because the Arabian Peninsula is one distinct geological and geographic entity that has suffered several natural disasters across the national boundaries. Most countries of this large peninsula do have government departments and other agencies responsible for implementing the state policy on disaster management and prevention, but in practice mostly they fall short of their responsibility whenever a natural hazard occurs in the region, for example, a lot of lives lost and multimillion dollar damage to property could have been minimized if there were clear guidelines for meeting events like 2009 flash floods of Jeddah in Saudi Arabia, and tropical cyclones Gonu in June 2007 and Phet in June 2010 in Oman among many natural disasters.

12.3.1 Possible Remediation of the Natural Disaster in the Arabian Peninsula

Geological hazards have struck this region during the past and there is enough scientific evidence to believe that this region is likely to be struck in the future also (Kumar 2009a, b). Construction activities are taking place at "break-neck" speed in most countries of the region specially the Arabian Gulf coastal regions. Some of the most famous constructions are the reclaimed "Palm Islands" communities in Dubai and few other similar communities being developed by Qatar and Bahrain. It is argued that the long-term sustainability of these reclaimed island communities is questionable. Thus, it is suggested that since the Arabian Gulf states have plenty of land and long coast line, new townships should be planned and developed around the onshore coastal regions. Onshore developments would be relatively safer than the communities developed on the reclaimed islands (Kumar 2009a, b).

The December 26, 2004 Sumatra earthquake followed by disastrous tsunamis killed over 250,000 people in Indonesia, Thailand, Sri Lanka and India and caused enormous damage to land and property. It forced the scientists, engineers, planners and authorities to think seriously about the issues of building code that meets seismicity of the region, protection from coastal flooding and planning construction of new townships that are relatively safer from such disasters (Harinarayana and Hirata 2005; Murty et al. 2007). Bou-Rabee and VanMarcke (2001) suggested that risk and vulnerability assessment against earthquakes and tsunamis are needed for all countries of the Arabian Gulf region.

There is a general belief among people that the Arabian Gulf countries are safe from natural disasters like earthquakes and tsunamis. Bou-Rabee and VanMarcke (2001) discussed threats from earthquakes and tsunamis in the Arabian Gulf region and emphasized that there is an urgent need for risk assessment of infrastructure systems, coastal developments and tall buildings. They also consider that building codes in the region currently lack seismic provisions despite enormous scientific evidence for such a threat. The long-term (100 ± 25 years) sustainability of the reclaimed island projects in the Arabian Gulf is questionable in view of possible geological hazards that include seismic, tsunami and/or tidal waves, and erosion due to surface and sub-marine currents. Movement of surface and submarine currents in the Arabian Gulf poses severe threat of erosion to these man-made islands. Constant stabilization effort will be required to sustain them.

12.3.2 Atmospheric Hazards and Probable Impact of Global Warming

Although tropical cyclones affecting the Arabian Peninsula and the Gulf are uncommon but they are becoming frequent during the past few years probably either due to the effect of global warming or some other climatic phenomenon. These can be severe and hazardous as demonstrated by widespread destruction and loss of life in Oman due to cyclone Gonu in June 2007. A wide range of issues related to global warming have been discussed by the Inter Governmental Panel on Climate Change (http://www.ipcc.ch) and various other scientific and environmental organizations. This issue is of vital importance to us because it could lead to far-reaching socio-economic and political changes in the world and potentially has disastrous consequences for all forms of life on Earth. In the Arabian Peninsula, over 90% of the area is suffering from some form of desertification and 44% of the area is severely degraded. Its plant biodiversity is low (~3,500 species), which is also getting reduced due mainly to overgrazing (Peacock et al. 2000). Goudie (2003), while discussing the impact of global warming on the geomorphology and hydrology of arid lands warns that *changes in vegetation cover, associated with decline of soil moisture availability, will increase the risk of soil erosion by wind, and may accelerate dust storm activity and cause sand dune activity to be reactivated.* He further states that, *arid coastlines will be subjected to sea-level change that will lead to transformations in the nature of such environments as deltas, estuaries and sabkhas.* As a result of global warming sea-levels are expected to rise at around 5 mm/yr. This could seriously impact low-lying coastlines such as the Sabkhas of UAE and due to the high degree of infrastructure development in such coastal environments this is a serious issue for cities such as Abu Dhabi (Goudie 2003). Thus it can be argued that such offshore townships built on the reclaimed islands could potentially be affected by inundation and erosion due to sea-level rise and increase in storm surge events.

12.4 Concluding Remarks

This paper provides an overview of geographical distribution, time framework and losses due to natural hazards that has impacted the Arabian Peninsula in the recent past. Among them the geological hazards discussed include earthquakes, tsunamis and tidal waves and volcanism. The atmospheric hazards discussed are tropical cyclones, storm surges and sand and dust storms. In addition droughts and desertification, and flash floods are also discussed in the light of "developmental" activities that include massive construction of infrastructures such as highways, ports, cities and recreation facilities in this region. A regional disaster management and prevention policy is needed to effectively tackle the consequences of natural disasters to minimize adverse impact on human population and the environment. The paper concludes by suggesting possible remediation measures against natural disasters in the Arabian Peninsula.

Acknowledgements I thank King Fahd University of Petroleum and Minerals, Saudi Arabia for permission to publish this paper.

References

Al-Amri AM (2005) Seismic source zones of the Arabian Peninsula and adjacent countries. Gulf Seismic Forum, United Arab Emirates University, Al-Ain, pp 2–3, Abstract

Al-Amri AM (2007) Characterization of seismic zones in the Arabian Peninsula. In: 7th meeting of the Saudi Society for Geosciences. King Saud University, Riyadh, p 56, Abstract

Al-Amri AM, Rodgers AJ, Alkhalifah TA (2007) Improving the level of seismic hazard parameters in Saudi Arabia using earthquake location and magnitude calibration, pp 1–25. http://faculty.ksu.edu.sa/5713/International%20Projects/AlAmri%20et%20al.pdf

Ali HA (1994) Wind regime of the Arabian Gulf. In: El-Baz F, Makharita RM (eds) The Gulf war and the environment. Gordon and Breach Science Publishers, New York, pp 31–48

Allison RJ et al (1997) Middle East and Arabia, chapter 22. In: Thomas DSG (ed) Arid zone geomorphology: processes, form and change in drylands. Wiley, Chichester, p 713

Alsharhan AS, Glennie KW, Whittle GL, Kendall CGSt.C (1998). Quaternary deserts and climate change. In: Proceedings of the international conference on quaternary deserts and climate change, Al Ain, 9–11 December 1995, A. A. Balkema Publishers, Rotterdam, p 621

Alsharhan AS, Wood WW, Goudie AS, Fowler A, Abdellatif EM (2003) Desertification in the third millennium. In: Proceedings of an international conference, Dubai, 12–15 February 2000. A. A. Balkema Publishers, Lisse, p 490

Ambraseys NN, Melville CP (1982) A history of Persian earthquakes. Cambridge University Press, New York

Ambraseys NN, Melville CP, Adams RD (2005) The seismicity of Egypt, Arabia and the Red Sea. Cambridge University Press, New York

Barnum BH, Winstead NS, Wesely J, Hakola A, Colarco PR, Toon OB, Ginoux P, Brooks G, Hasselbarth L, Toth B (2004) Forecasting dust storms using the CARMA-dust model and MM5 weather data. Environ Model Softw 19:129–140

Beer T (2010) Geophysical hazards: minimizing risk, maximizing awareness. Springer, Heidelberg, p 262

Bhaskaran PK, Dube SK, Murty TS, Gangopadhyay A, Chaudhury A, Rao AD (2007) Tsunami travel time atlas for the Indian Ocean. In: Murty TS, Aswathanarayana U, Nirupama N (eds) The Indian Ocean tsunami. Taylor and Francis, London, pp 273–292

Bou-Rabee F, VanMarcke V (2001) Seismic vulnerability of Kuwait and other Arabian Gulf countries: information base and research needs. Soil Dyn Earthq Eng 21:181–186

Chadha RK (2007) Tsunamigenic sources in the Indian Ocean: factors and impact on the Indian landmass. In: Murty TS, Aswathanarayana U, Nirupama N (eds) The Indian Ocean tsunami. Taylor and Francis, London, pp 33–48

Dimri VP, Srivastava K (2007) Modelling techniques for understanding the Indian Ocean tsunami propagation. In: Murty TS, Aswathanarayana U, Nirupama N (eds) The Indian Ocean tsunami. Taylor and Francis, London, pp 123–130

Dube SK, Rao AD, Sinha PC, Murty TS, Bahulayan N (1997) Storm surge in the Bay of Bengal and Arabian sea: the problem and its prediction. Mausam 48:283–304

El-Isa ZH, Al-Shanti A (1989) Seismicity and tectonics of the Red Sea and western Arabia. Geophys J Int 97(3):449–457

El-Sabh MI, Murty TS (1989) Storm surges in the Arabian Gulf. Nat Hazard 1:371–385

Fournier M, Lepvrier C, Razin P, Jolivet L (2006) Late Cretaceous to Paleogene post-obduction extension and subsequent Neogene compression in the Oman Mountains. GeoArabia 11:17–39

Glenie KW, Singhvi AK (2002) Event stratigraphy, paleoenvironment and chronology of SE Arabian deserts. Quat Sci Rev 21:853–869

Goudie AS (2003) The impacts of global warming on the geomorphology of arid lands. In: Al-Sharhan AS, Wood WW, Goudie AS, Fowler A, Abdellatif EM (eds) Desertification in the third millennium. Balkema, Lisse, pp 13–19

Grainger D (2007) The geological evolution of Saudi Arabia: a voyage through space and time. Saudi Geological Survey, Jeddah, p 264

Harinarayana T, Hirata N (2005) Destructive earthquake and disastrous tsunami in the Indian Ocean, what next? Gondwana Res 8:246–257

Harrigan P (2006) Volcanic Arabia. Saudi Aramco World 57:2–13. http://www.saudiaramcoworld.com/issue/200602/volcanic.arabia.htm

Heidarzadeh M, Pirooz MD, Zaker NH, Yalciner AC, Mokhtari M, Esmaeily A (2008) Historical tsunami in the Makran subduction zone off southern coasts of Iran and Pakistan and results of numerical modeling. Ocean Eng 35:774–786

Heidarzadeh M, Pirooz MD, Zaker NH, Yalciner AC (2009) Preliminary examination of the tsunami hazards associated with the Makran subduction zone at the northwestern Indian Ocean. Nat Hazard 48:229–243

Jordan BR (2008) Tsunamis of the Arabian Peninsula, a guide of historic events. Sci Tsunami Hazard 27:31–46

Jordan BR, Baker H, Howari F (2006) Tsunami hazards along the coasts of the United Arab Emirates. Dubai Municipality Publication

Khan MA, Kumar A (2009) Impact of "urban development" on mangrove forests along the west coast of the Arabian Gulf. e-journal Earth Sci India 2:159–173

Khan MA, Kumar A, Muqtadir A (2010) Distribution of mangroves along the Red Sea coast of the Arabian Peninsula: Part-2. The southern coast of western Saudi Arabia. e-journal Earth Sci India 3:154–162

Kumar A (2009a) Reclaimed islands and new offshore townships in the Arabian Gulf: potential natural hazards. Curr Sci 96:480–485

Kumar A (2009b) History of earthquakes and tsunamis in the Arabian Gulf: potential hazard for reclaimed island communities. In: WCCE-ECCE-TCCE joint conference: earthquakes and tsunamis, June 22–24, 2009, Istanbul. Proceeding volume, pp 1–7

Kumar A, Alam SA (2010) "Shock Event", an impact phenomenon observed in water wells around the Arabian Gulf coastal city Dammam, Saudi Arabia: possible relationship with Sumatra tsunami event of December 26, 2004. Nat Hazard 53:407–413

Kumar A, Khan MA, Muqtadir A (2010) Distribution of mangroves along the Red Sea coast of the Arabian Peninsula: Part-I: the northern coast of western Saudi Arabia. e-journal Earth Sci India 3(28)

McKee ED (1979) A study of global sand seas. United States Geological Survey Professional Paper 1052, p 429
Middleton NJ (1986) Dust storms in the middle East. J Arid Environ 10:83–96
Middleton N (1997) Desert dust. In: Thomas DSG (ed) Arid zone geomorphology; processes, form and change in drylands, 2nd edn. Wiley, London, pp 413–436
Mokhtari M, Fard IA, Hessami K (2008) Structural elements of the Makran region, Oman sea and their potential relevance to tsunamigenesis. Nat Hazard 47:185–199
Murty TS (1999) Strom surges in the marginal seas of the North Indian Ocean. In: Proceedings of the WMO/UNESCO sub-forum on science and technology in support of natural disaster reduction, Geneva, July 6–8, 1999, World Meteorological Organization (WMO-No. 914), pp 130–139
Murty TS, El-Sabh MI (1984) Cyclones and storm surges in the Arabian Sea: a brief review. Deep-Sea Res 31:665–670
Murty TS, Bapat A, Prasad V (1999) Tsunamis on the coastlines of India. Sci Tsunami Hazard 17:167–172
Murty TS, Aswathanarayana U, Nirupama N (2007) The Indian Ocean tsunami. Taylor and Francis, London, pp 1–491
Pararas-Carayan G (2006) The potential of tsunami generation along the Makran subduction zone in the northern Arabian Sea: Case study: the earthquake and tsunami of November 28, 1945. Sci Tsunami Hazard 24(5):358–384
Peacock JM, Ferguson ME, McCann IR, Alhadrami GA, Karnik R, Saleh A (2000) Desert forages of the Arabian Peninsula – the conservation and utilization of biodiversity for sustainable animal production in the United Arab Emirates. In: Al-Sharhan AS, Wood WW, Goudie AS, Fowler A, Abdellatif EM (eds) Desertification in the third millennium. Balkema, Lisse, pp 131–138
Petrovski JT (2005) Seismic risk of tall buildings and structures caused by distant earthquakes. Gulf Seismic Forum, United Arab Emirates University, Al-Ain, pp 35–40, Abstract
Pye K (1987) Aeolian dust and dust deposits. Academic, London
Rastogi BK (2007) A historical account of the earthquakes and tsunamis in the Indian Ocean. In: Murty TS, Aswathanarayana U, Nirupama N (eds) The Indian Ocean tsunami. Taylor and Francis, London, pp 3–18
Rastogi BK, Jaiswal RK (2006) A catalogue of tsunamis in the Indian Ocean. Sci Tsunami Hazard 25:128–143
Rodgers A (2005) Ground motion in the gulf region from large earthquakes in Zagros Mountains. Gulf Seismic Forum, United Arab Emirates University, Al-Ain, p 41, Abstract
Seach J (2010) The Red Sea volcanoes. http://www.volcanolive.com/redsea.html
Shehata WM, Bader T, Harari ZY (1990) Principal geological hazards along the Arabian Gulf coast. J King Abdulaziz Univ 3:289–301, Earth Science, Special issue: first Saudi symposium on Earth Sciences, Jeddah, 1989
Siebert L, Simkin T (2002) Volcanoes of the World: an illustrated catalog of holocene volcanoes and their eruptions. Smithsonian Institution. Global Volcanism Program Digital Information Series, GVP-3. http://www.volcano.si.edu/world/
Simkin T, Siebert L (1994) Volcanoes of the world, 2nd edn. Geoscience press in association with the Smithsonian Institution Global Volcanism program, Tucson, p 368
Spiess A (2008) Developing adaptive capacity for responding to environmental change in the Arabian Gulf states: uncertainties to linking ecosystem conservation, sustainable development and society in authoritarian rentier economies. Glob Planet Chang 64:244–252
Squires VR (2003) Desertification, climate change and the world's drylands. In: Al-Sharhan AS, Wood WW, Goudie AS, Fowler A, Abdellatif EM (eds) Desertification in the third millennium. Balkema, Lisse, pp 13–20
Tegen I, Fung I (1994) Modeling of mineral dust in the atmosphere: sources, transport and optical thickness. J Geophys Res 99:22897–22914

Thomas DSG (1997) Arid zone geomorphology: processes, form and change in drylands, 2nd edn. Wiley, London, p 713
Thomas DSG (2003) Into the third millennium: the role of stakeholder groups in reducing desertification. In: Al-Sharhan AS, Wood WW, Goudie AS, Fowler A, Abdellatif EM (eds) Desertification in the third millennium. Balkema, Lisse, pp 3–12
Verstraete M, Brink A, Scholes R, Smith MS (2008) Climate change and desertification. Glob Planet Change 64:105–254
Washington R, Todd M, Middleton NJ, Goudie AS (2003) Dust-storm source areas determined by the total ozone monitoring spectrometer and surface observations. Ann Assoc Am Geogr 93:297–313
Williams CT (2005) Development and seismic stability in coastal zone. Gulf Seismic Forum, UAE University, Al-Ain, p 57, Abstract

Chapter 13
Influence of Young's Modulus and Poisson's Ratio on the Displacement Around a Circular Tunnel

A.K. Verma and T.N. Singh

13.1 Introduction

Underground structures in rock or soil are excavated for many useful purposes e.g. hydro-power generation, transportation, storage, communication, etc. Underground excavations have geometrical shape and size such as caverns, drifts, cross-cuts, shafts and tunnels etc. depending upon the need. The design of any underground excavation primarily depends on three basic criteria viz. stability, economy and safety. Stability of underground structure depends on many parameters such as mechanical properties for rock mass, geological condition of the site, in-situ stresses as well as geometry of openings. From analytical point of view the safety and stability of structures depend on distribution of stress as well as its load bearing capacities against the excavation patterns. The study of stress distribution and displacement plays a crucial role in determining the stability of underground structures. Also, prediction of in situ stresses and deformation modulii by back-analysis of tunnel convergence measurement and of 'Ground Reaction Curve' is essential for design of support system in tunnels (Obert and Duvall 1967; Hoek and Brown 1980; Brady and Brown 1985).

When, an underground structure is excavated in virgin rock, isostasy in the rock mass, achieved over geologic time, is disrupted and the forces are redistributed initially in haphazard manner. If the stresses redistributed around an opening exceed the bearing capacity of the rock mass, the failure of the rock mass adjusted to the excavated boundary may lead to instability. Numerical analyses for studying the stability have always enthralled researchers because of its flexibility in changing various parameters. Numerical modeling utilizes the computational power of computers to create complex models based on simple theoretical models of rock mass behavior. The purpose of this study is to analyze stresses and deformation around circular tunnel by varying elastic parameters such as Young's modulus and

A.K. Verma (✉) • T.N. Singh
Department of Earth Science, Indian Institute of Technology, Bombay 400076, India
e-mail: amverma@iitb.ac.in

R. Sinha and R. Ravindra (eds.), *Earth System Processes and Disaster Management*, Society of Earth Scientists Series 1, DOI 10.1007/978-3-642-28845-6_13,

Poisson's ratio along the longitudinal axis of tunnel under hydrostatic stress field. The numerical simulation have been executed for studying various underground analysis such as tunnel convergence (Pellet and Verma 2006; Sitharam et al. 2005; Verma and Singh 2010; Singh and Verma 2007, 2009), analyzing rock-support interaction, determining excavation disturbance zone, evaluating tunnel face instability, etc involving different rock types (Unlu and Gercek 2003; Pelli et al. 1991).

For the present study, a three dimensional numerical code, FLAC^{3D} has been employed for the analysis of deformation and stress distribution in a circular tunnel. Due to the three-dimensional stress distribution at tunnel faces, straightforward application of two-dimensional numerical analysis to the design of tunnel support systems may be inaccurate. The 2-D numerical modeling does not tell about deformation, stress distribution and other parameters along the longitudinal axis of tunnel. Most of two-dimensional numerical formulations for excavation analysis assume plane strain conditions when one of the principal components of the in situ stresses is acting parallel to the longitudinal axis of the opening (Hoek and Brown 1980; Panet 1993). Three dimensional analyses can provide useful information at critical places such as intersection of tunnels, excavation face, junctions etc.

It has been observed that radial deformation decreases with increase in Young's modulus (Unlu and Gercek 2003). The empirical relations relating radial deformation and distance along the longitudinal axis of tunnel considering the effect of elastic parameters (E and ν) has been provided by numerical FLAC^{3D} analysis. Many researchers (Unlu and Gercek 2003; Corbetta et al. 1991; Panet 1995; Panet and Guenot 1982; Guilloux et al. 1996; Sulem et al. 1987) have provided empirical relation for deformation around circular tunnels as function of longitudinal distance. They did not, however, take into consideration the effect of elastic parameters. Unlu and Gercek (2003) made an attempt to study the effect of Poisson's ratio on the deformation in underground excavations.

The critical distance from the tunnel face decreases as Young's modulus increases. Similarly, radial deformation increases with increase in Poisson's ratio for all observational point along the longitudinal axis of tunnel. Similar behavior has been observed by Unlu and Gercek (2003). The deformation obtained from the numerical FLAC^{3D} calculation, along the longitudinal direction for circular tunnel have been compared with work done by other researchers (Unlu and Gercek 2003; Carranza-Torres and Fairhurst 2000).

13.2 Analytical Solutions

Many researchers have studied the deformation and stress distribution around the circular opening in past using different numerical and analytical methods. For studying stresses and displacement in preliminary stage a simplified assumption that tunnel of circular cross-section has been dealt (Hoek and Brown 1980; Brady and Brown 1985; Jaeger and Cook 1976). Elaborate study has been provided by Obert and Duvall (1967), Hoek and Brown (1980), and Brady and Brown (1985) primarily by analytical study for underground circular opening. Brady and Brown (1985) have provided analytical solution for radial displacement around a circular opening given as:

$$u_r = -\frac{P_v R^2 (1+v)}{2Er} \times \left\{ (1+k) - (1-k)\left[4(1-v) - \frac{R^2}{r^2}\right] \cos 2\,\theta \right\},$$

$$u_\theta = -\frac{P_v R^2 (1+v)}{2Er} \left\{ (1-k)\left[2(1-2v) + \frac{R^2}{r^2}\right] \sin 2\,\theta \right\}$$

Where Pv and Ph are vertical and horizontal in situ stress, ur and u (theta) are radial and tangential displacements, k is ratio of horizontal to vertical in situ stress component, r is distance along the radial direction and theta is angle from the horizontal.

For hydrostatic stress field k = 0, and $P_v = P_h = P_o$. Thus equations become:

$$u_r = -\frac{P_o R^2 (1+v)}{Er} \quad (13.1)$$

and

$$u_{r\infty} = \frac{P_o R (1+v)}{E} \quad (13.2)$$

On the other hand, Panet and Guenot (1982), Panet (1995), Corbetta et al. (1991), Carranza-Torres and Fairhurst (2000) and Unlu and Gercek (2003) has provided numerical solution to deformation around circular opening using finite element and finite difference methods. Most of them have used hydrostatic field conditions for better understanding of the problem and to facilitate the use of axisymmetric solution. These equations are based on general hyperbolic relation between radial deformation and distance along the longitudinal axis. None other than Unlu and Gercek (2003) have proposed an equation relating deformation around tunnel boundary showing the dependency on Poisson's ratio with numerical $FLAC^{3D}$ modeling. They considered Poisson's ratio in all of its range i.e. from 0.05 to 0.45 which is a general variation of Poisson's ratio. On the other hand, studying the effect of Young's modulus in range 1.5–29 GPa is a realistic consideration with respect to common sedimentary rock type especially for the Indian rocks. Rocks falling in this Young's modulus range include sandstones, schist, gneiss, quartzite, phyllite etc. The following equations have been proposed by different researchers:

$$\frac{u_r}{u_{r\infty}} = 0.28 + 0.72\left[1 - \left(\frac{0.84}{0.84 + x/R}\right)^2\right] \tag{13.3}$$

by Panet and Guenot (1982)

$$\frac{u_r}{u_{r\infty}} = 0.25 + 0.75\left[1 - \left(\frac{0.75}{0.75 + x/R}\right)^2\right] \tag{13.4}$$

by Panet revised (1995)

Some researchers have suggested exponential equations for the normalized radial as deformation such as:

$$\frac{u_r}{u_{r\infty}} = 0.29 + 0.71\{1 - \exp[-1.5(x/R)^{0.7}]\} \tag{13.5}$$

by Corbetta et al. (1991) and

$$\frac{u_r}{u_{r\infty}}\left[1 + \exp\left(\frac{-x/R}{1.10}\right)\right]^{-1.7} \tag{13.6}$$

by Carranza-Torres and Fairhurst (2000)

$$\mathrm{u_r} = \mathrm{u_{r\infty}}\Big((0.22\mathrm{v} + 0.19) + (-0.22\mathrm{v} + 0.81)\left\{1 - ((0.39 + 0.65)/(0.39 + 0.65 + \mathrm{x/R}))^2\right\}\Big) \tag{13.6a}$$

by Unlu and Gercek (2003)

Where u_r is distance from tunnel face, $u_{r\infty}$ is contact deformation at critical distance from the tunnel face and x is distance from tunnel face taken positive behind the face and negative ahead of face (only for (13.6)).

They all have found that the ratio of the normalized value of pre-deformation is a function of ratio of distance from the face divided by radius of tunnel. These equations are valid for elastic behavior of rock mass.

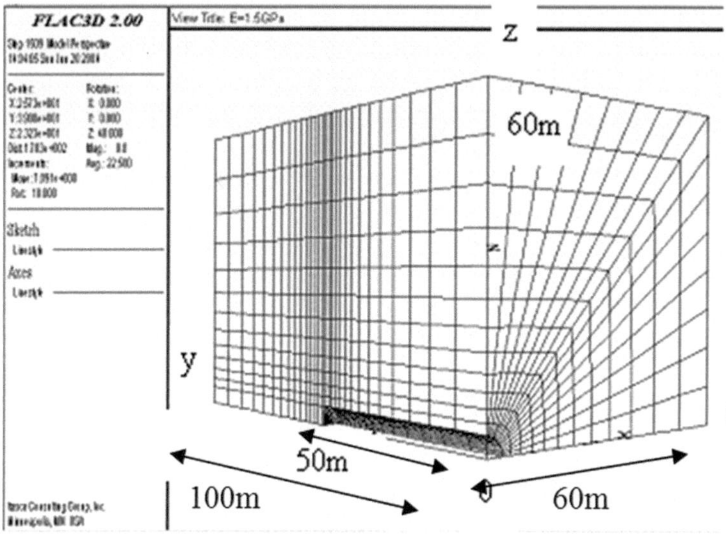

Fig. 13.1 Numerical model for the simulation of circular tunnel (longitudinal axis along the Y axis)

13.3 Numerical Modeling

13.3.1 Model Geometry

The grid of the model geometry is shown in Fig. 13.1. The radial brick elements are used in the model geometry. The area of observation is chosen mainly at the tunnel face and at the radial boundary of the tunnel along the longitudinal axis of tunnel. After a parametric study on the variation of tunnel diameter (from 2 to 6 m), tunnel diameter has been selected as 3 m. The size in the model grid is reduced successively in the ratio of 0.85 from the radial and model boundaries towards the tunnel face to get more accurate deformations. As the grid becomes smaller and smaller, the monitoring point lies on the node of the grid and the deformations are calculated accurately. The value of 0.85 has been taken after a parametric study where ratio was varied from 0.5 to 1.0. It is important to note that side of square cross-sectional area of tunnel is taken 20 times of tunnel diameter to avoid effect of boundary on simulation (Itasca 1997). At that distance, effect of excavation is assumed to be negligible. Total numbers of zones and grid points are 6,000 and 6,867 respectively.

13.3.2 Boundary Conditions

The symmetry of the circular tunnel in geometry and boundary conditions helped in considering only one quarter of tunnel and outer rock mass is taken for the numerical simulation. On the other hand, if the weight of the rock in the yield zone around the tunnel periphery would have included in the analysis, this simplifying symmetry would have lost (Hoek and Brown 1980; Unlu and Gercek 2003). The reduction in number of zones due to symmetry account for increase in speed of simulation because as the number of zone increases, simulation speed decreases. The displacement constraints were given in x, y and z direction at the planes $x = 0$, $y = 0$, and $z = 0$ respectively; all passing though the origin. The in-situ stress of 30 MPa was initialized in each zone of numerical FLAC^{3D} model. The hydrostatic pressure $P_0 = 30$ MPa was applied at planes $x = 60$ m and $y = 100$ m. The vertical component $\sigma_{zz} = 30$ MPa is also applied at plane $z = 60$ m. These constraints are normal to symmetry planes, and therefore, displacement observed on this plane may be under some error (Itasca 1997). To avoid this effect, points of observation are selected on the plane along the longitudinal axes making an angle of 45° with x-axis (Unlu and Gercek 2003). The displacement histories in x and z direction at these selected points was recorded during simulation process. A FISH function named 'raddisp' was also incorporated in the numerical code to calculate the radial displacement on this plane. Radial displacement is resultant displacement along x and z axis i.e. $u_r = \sqrt{(u_x^2 + u_z^2)}$.

13.3.3 Properties of Model

The tunnel is assumed to be driven in a homogeneous, isotropic, linear elastic rock mass. In this study, effect of Young's modulus has been studied by varying it from 1.5 to 29 GPa at constant Poisson's ratio, $\nu = 0.25$. These rock mass properties account for the soft to medium hard rock condition.

The effect of Poisson's ratio has been studied with Young's modulus 1.5 GPa and Poisson's ratio varying in all of its range i.e. from 0.05 to 0.45. After initializing material properties and boundary conditions to geometric model, model was simulated. Histories of displacement along x and z axis are recorded up to 50 m behind the tunnel face and ahead of the tunnel face up to 4 m along the radial boundary of tunnel. The unbalance force history in the model was also recorded and shows equilibrium condition of rock mass strata (Fig. 13.2).

13.4 Results and Discussion

13.4.1 Effect of Young's Modulus on Deformation in Advancing Tunnel

13.4.1.1 Deformation Behind the Tunnel Face

The variation of radial boundary deformation along the longitudinal axis of the simulated tunnel located in hydrostatic in situ stress field at different Young's

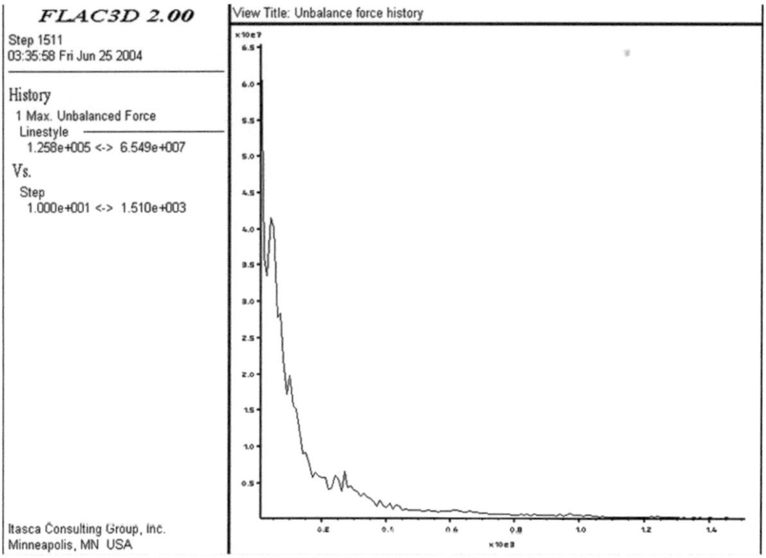

Fig. 13.2 Unbalance force history of the simulated model

modulus varying from 1.5 to 29 GPa is shown in Fig. 13.3. Along the abscissa axis in the graph, origin is taken at the tunnel face and it is increasing towards the region located behind the tunnel face. It can be deduced from Fig. 13.3 that, for a particular Young's modulus, towards the region behind the face, there is tendency to increase in radial boundary deformation. It may be due to reason that, behind the face area, unsupported span of tunnel increases and hence deformation as well. Later on, radial deformation becomes constant far from the tunnel face. It may be suggested that, as tunnel face advances, there is sufficient time available for redistribution of stresses of the rock mass behind the face. Similar behavior has been observed thought analytical study and field investigation by many researchers (Abel and Lee 1973) ahead of tunnel face.

The empirical equation for the above curve (Fig. 13.3) between length along longitudinal axis of tunnel and radial deformation at varying Young's modulus has been purposed by curve fitting. The best fit of simulated data variables was found to be with sigmoid function. Other researchers have also chosen similar curve types to observe the radial deformation. Also, there is no closed-form solution for study of Young's modulus on deformation around circular opening to best of authors knowledge derived empirically. The best correlation has been obtained by sigmoid function and equation is given below:

$$u_r = \mathbf{A_2} + (\mathbf{A_1} - \mathbf{A_2})/(\mathbf{1} + \mathbf{t}(\mathbf{x}/\mathbf{R})^{\mathbf{p}}) \qquad (13.7)$$

The coefficient of correlation coming out to be more than 0.999 in all cases except at E = 15 GPa which is 0.998. The boundary conditions for this equation are:

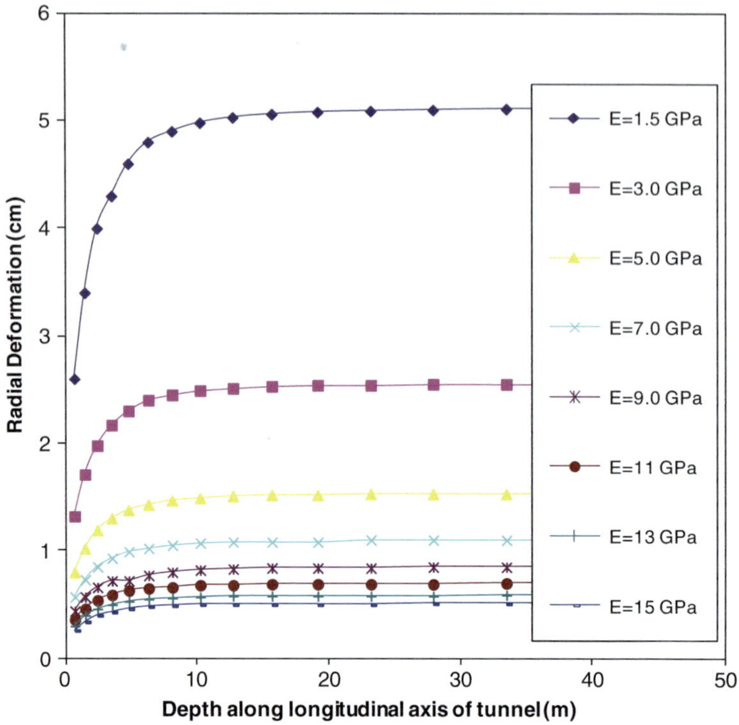

Fig. 13.3 Variation of radial boundary deformation along the longitudinal axis of circular tunnel located in hydrostatic in situ stress field at different Young's modules (1.5–15 GPa)

at $x = 0$, $u_r = A_1$ and at $x \to \infty$, $u_r \to A_2$. So, A_1 and A_2 are simply radial deformations at tunnel face and at critical distance respectively. The critical distance is distance from tunnel face beyond which deformations becomes constant. The values of A_1 and A_2 were found from numerical and statistical analysis and both are showing quite good correlation. The statistical parameters A_1 and A_2 have been found to be almost inversely dependent on Young's modulus (with correlation coefficient 0.98 and 0.999 respectively).

$$\mathbf{A_1 = 3.2E^{-1.04}} \tag{13.8}$$

and

$$\mathbf{A_2 = 7.75E^{-1.0}} \tag{13.9}$$

The parameters t and p in the proposed equation (13.7) are found to be constant having approximate values 2.33 and 1.5 respectively.

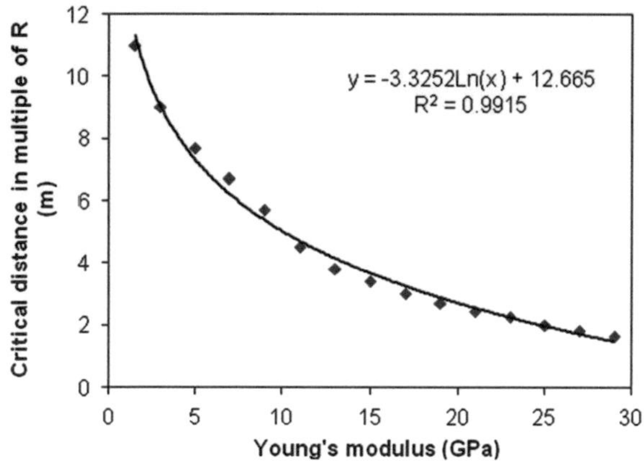

Fig. 13.4 Plot of variation of critical distance (in multiple of tunnel radius) versus Young's modulus

Figure 13.4 displays the variation of critical distance (in multiple of tunnel radius, R = 3 m) with Young's modulus. As Young's modulus is increased, the critical distance shifted near the tunnel face. It may be due to reason that, more elastic material would come to its original state quicker. The best fit between Young's modulus and critical distance is shown in Fig. 13.4 and empirical relation between them as suggested as: (coefficient of correlation $r^2 = 0.9915$).

$$\mathbf{x = -3.33}\ \boldsymbol{ln}\ (\mathbf{E}) + \mathbf{12.67} \tag{13.10}$$

Where x is critical distance from tunnel face and E is Young's modulus. This indicates increase in stability of rock strata in excavated region at higher Young's modulus. Bieniawski (1978) (for E > 10 GPa) and Serafim and Pereira (1983) (for E < 10 GPa) developed the empirical relation $E_m = 2(RMR) - 100$ and $E_m = 10^{(RMR - 10)/40}$ respectively based upon the field test results at sites throughout the world (Itasca 1997). So, the Young's modulus is dependent upon the condition of rock mass which can be indirectly inferred from RMR values. Now according to Hoek and Brown (1980), the unsupported span in underground excavation is function of Q (quality index of rock mass) i.e. span = $2.ESR.Q^{0.4}$. So, this implies that unsupported span depends on Young's modulus. Clearly, for better rock quality, unsupported span would be more.

Figure 13.5 displays the variation of radial boundary displacement with Young's modulus at different locations behind the circular tunnel face. The points of observation behind the tunnel face are taken at gradually increasing distance. This is because zones were selected in the geometry such that, understanding the mechanism of rock behavior near the face can be more precise (Itasca 1997).

It is further noted that as the Young's modulus is increased from 1.5 to 3 GPa, the radial deformation reduced from 2.6 to 1.32 cm. This implies that for two

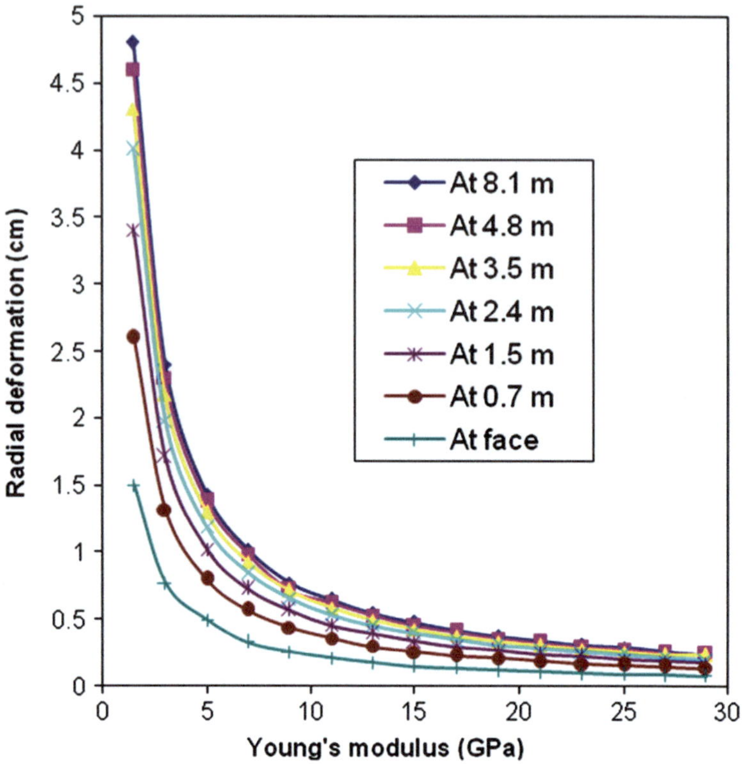

Fig. 13.5 Variation of radial boundary displacement with Young's modulus at different locations behind the circular tunnel face in hydrostatic stress field

times increase in Young's modulus caused radial boundary displacement to become half. This behavior continues all along the curve and also at different locations along the tunnel axis. Observation in the Fig. 13.5 beyond 8.1 m is not displayed because difference in deformation is stagnant after that. The following proposed empirical relation from numerical data has been established by curve fitting (coefficient of correlation is more than 0.99 at all observational points):

$$\mathbf{u_r = mE^{-n}} \tag{13.11}$$

Where n is exponent and found to vary between from 1.0014 to 1.0039 (almost equal to 1) and m from 2.33 to 7.20. So, it can be concluded from the numerical results that radial displacement is inversely proportional to Young's modulus. The analytical study by Brady and Brown (1985) (13.1) shows the similar behavior.

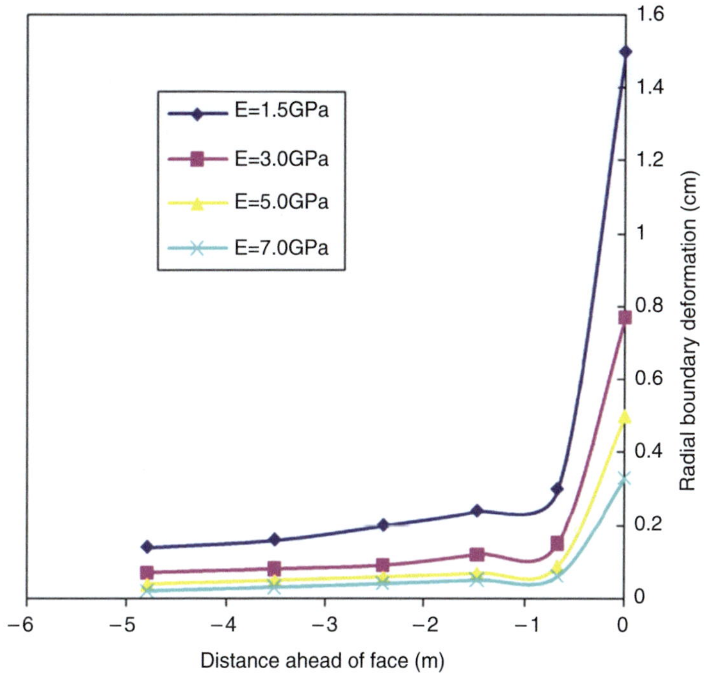

Fig. 13.6 Radial boundary deformation along the longitudinal axis of a circular tunnel ahead of face at different Young's modulus

13.4.1.2 Deformation Ahead of Tunnel Face

The variation of radial boundary deformation with distance along the longitudinal axis ahead of face at various Young's modulus is shown in Fig. 13.6. The radial boundary displacement histories were recorded at five points ahead of tunnel face at different Young's modulus. It can be seen that, just after the tunnel face, there is sudden decrease in radial displacement. It decreases further into the rock mass. As the Young's modulus increase, the value of radial deformation decreases and tends to become zero away from the tunnel face.

There is a large difference in deformation at Young's modulus 1.5 and 3.0 GPa. As Young's modulus increases further, the difference in deformation becomes lesser. So, at higher values of the Young's modulus, sensitivity of rock mass towards deformation decreases or in other words, failure of rock mass becomes less prone.

Deformation histories along the length of tunnel ahead of face records shows its variation at different Young's modulus as shown in Fig. 13.7. It can be observed that, there is considerable change in radial deformation at the face and ahead of the face in the rock mass. Similar behavior was observed in the Fig. 13.6 also. From the

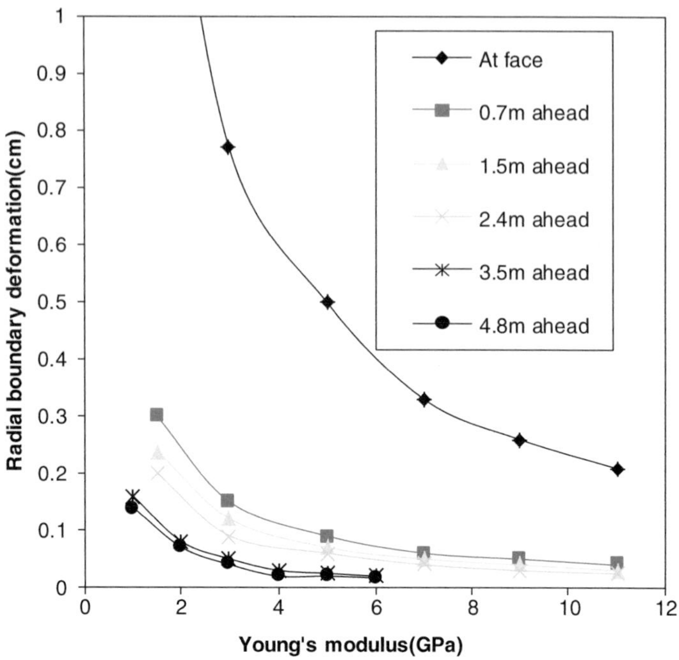

Fig. 13.7 Variation of radial boundary deformation with Young's modulus at different location along the longitudinal axis

numerical analysis it was found that, deformation becomes negligible at 2.4 m (at 0.8R) and 4.8 m (at 1.6R) ahead of tunnel face for Young's modulus of approximately 13 and 7 GPa respectively.

13.4.1.3 Effect of Young's Modulus on Deformation and Its Comparison

Figure 13.8a–c exhibit the comparison of proposed equation (13.10) based on simulated FLAC^{3D} model and proposed equation by Panet (1995), Carranza-Torres and Fairhurst (2000) and Unlu and Gercek (2003) given by (13.4), (13.6), and (13.6a) respectively at Young's modulus (E) equal to 1.5 GPa, GPa and 15 GPa. There is quite good correlation in three equation except (13.6) by Fairhurst. The equation proposed by Fairhurst has deviated from analytical solution because of positive sign of exponential function and its negative power.

In all these three cases, Poisson's ratio is 0.25 and putting v = 0.25 in (13.6a) makes it very close to (13.4). The advantage by the proposed equation (13.10) is that, it provides the affect of extra parameter i.e. Young's modulus in studying the radial deformation in underground excavations.

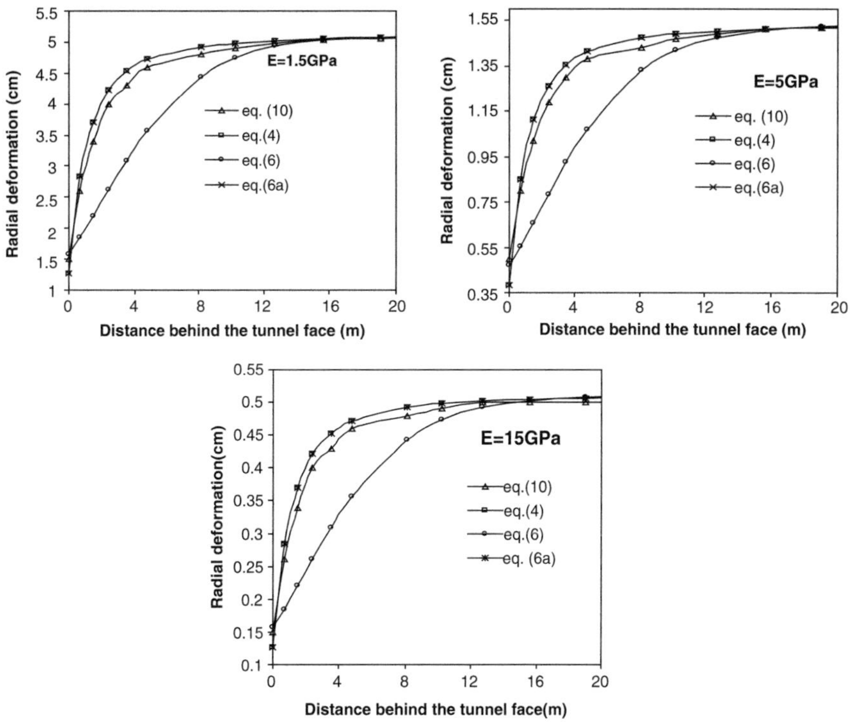

Fig. 13.8 (**a–c**) Comparison of simulated and previous proposed equations

13.4.2 Effect of Poisson's Ratio on the Deformation in Advancing Tunnel

13.4.2.1 Deformation Behind the Tunnel Face

Results of numerical analyses demonstrate that the radial deformation in the tunnel, as expected, is affected by the Poisson's ratio of the surrounding medium. Previously, Sakurai (1978), Niwa et al. (1979), Guilloux et al. (1996) and Unlu and Gercek (2003) have studied the effect of Poisson's ratio on radial deformation. Figure 13.9 shows the variation of radial deformation along the longitudinal direction with Poisson's ratio at different distance from the tunnel face. It can be observed from the curve that as the Poisson's ratio increases there is linear increase in radial deformation at all observational points. The empirical relation has been proposed between radial deformation and Poisson's ratio (with a coefficient of determination i.e. r^2 between 0.9985 and 0.9962) is obtained.

$$\mathbf{u_r} = \mathbf{m}\nu + \mathbf{c} \tag{13.12}$$

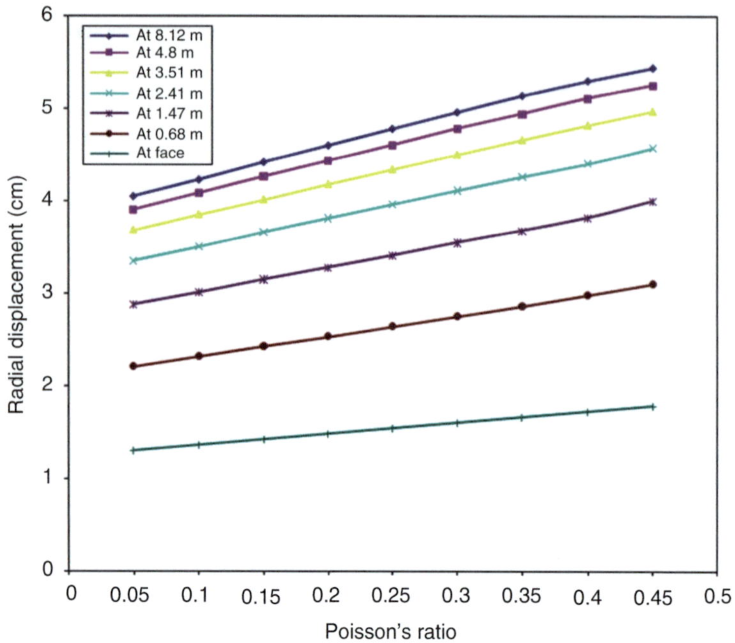

Fig. 13.9 Variation of boundary radial deformation with Poisson's ratio at different location along the longitudinal direction

Where m and c are constants varying from 1.2 to 4.0 and 1.24 to 4.10 respectively as Poisson's ratio increases from 0.05 to 0.45.

It is clear from (13.12) that slope of line is positive. It means that as Poisson's ratio increases, radial deformation also increases. The distance between successive lines is decreasing geometrically while moving away from tunnel face. This implies that there is large difference in deformation near the face. It is also observed that the slope of lines in the Fig. 13.9 is increases while moving away from tunnel face. This shows that, change of Poisson's ratio has less effect on radial deformation near the tunnel face. This may be due to reason that, increase in Poisson's ratio causes relatively more lateral displacement for same amount of axial displacement.

Figure 13.10 displays the variation of radial deformation behind the tunnel face at varying Poisson's ratio. It can be emphasized that, near the tunnel face, deformation increase linear along longitudinal axis and later becomes constant. Also, for same increase of Poisson's ratio, deformation increases by same amount through all range of Poisson's ratio. It can also be noted that, critical distance is same for every value of Poisson's ratio. This implies that critical distance is independent of Poisson's ratio.

The empirical equation for the curve (Fig. 13.10) between length along longitudinal axis of tunnel and radial deformation at varying Poisson's ratio has been

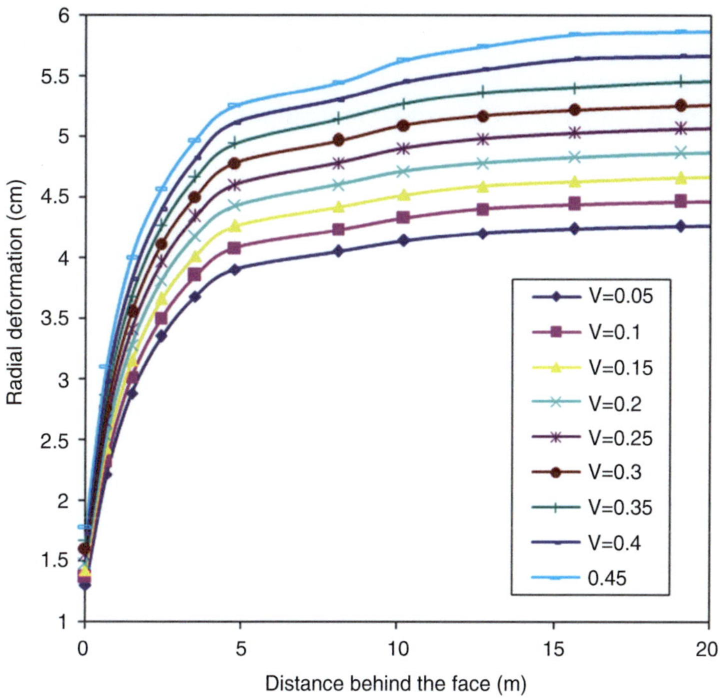

Fig. 13.10 Variation of boundary radial deformation along the longitudinal axis at different Poisson's ratio

purposed by curve fitting (with correlation coefficient $r^2 > 0.999$ for $0.05 < v < 0.45$). The proposed equation is again found to be best fit by sigmoid function curve. Similar functions have been used by other researchers.

$$u_r = \mathbf{B_2 + (B_1 - B_2)/(1 + z(x/R)^n)} \tag{13.13}$$

Where, B_1, B2, z and n are statistical constants. The boundary conditions for this equation are: at $x = 0$, $u_r = B_1$ and at $x \to \infty$, $u_r \to B_2$. So, B_1 and B_2 are simply radial deformations at tunnel face and at critical distance. The values of B_1 and B_2 were found from numerical and statistical analysis and both found to have quite good correlation. The statistical parameters B_1 and B_2 have been found to be linearly dependent on Poisson's ratio (with correlation coefficient 0.9999 and 1.0 respectively).

$$\mathbf{B_1 = 4.02}v\mathbf{ + 4.1} \tag{13.14}$$

and

$$\mathbf{B_2 = 1.2\nu + 1.24} \tag{13.15}$$

The parameters z and n in the proposed equation (13.15) are found to be constant having approximate values 2.33 and 1.5 respectively.

13.4.2.2 Ahead of Tunnel Face

The variation of radial deformation at different Poisson's ratio ahead of face is shown in Fig. 13.10. It can be clearly seen that, deformation curve at different Poisson's ratio overlaps for distance less than 2.5 m. So, there seems to be no effect of Poisson's ratio on the deformation ahead of face except at the face. It seems that for distance greater than 2.5 m the deformation increases with the increase in Poisson's ratio.

These results are in agreement contradiction with the results obtained from Unlu and Gercek (2003). However, the displacement magnitude noted matches with the empirical relation given by other researchers (13.3–13.6), which is displayed in next section.

Figure 13.11a–c shows the comparison of empirical equations by Panet (1995), Unlu and Gercek (2003) and equation proposed by numerical FLAC^{3D} model and their equations are (13.4), (13.6), (13.6a) and (13.13) respectively.

13.4.2.3 Effect of Poisson's Ratio on Deformation

Figure 13.11a–c shows the comparison of empirical equations by Panet (1995), Carranza-Torres and Fairhurst (2000), Unlu and Gercek (2003) and equation proposed by numerical FLAC^{3D} model and their equations are (13.4), (13.6), (13.6a) and (13.13) respectively.

Their comparison is almost similar as that for effect of Young's modulus. The results obtained by (13.6) are in disagreement with results with other three equation. The results of proposed equation are in close relation with empirical equations (13.4), (13.6) and (13.6a).

13.4.3 Analysis of Stability from Volumetric Strain, Radial Deformation and Displacement Vector Distribution in the Rock Mass

Figure 13.12 displays the contour of total displacement vectors in and around the circular tunnel in hydrostatic stress field in the geometric simulated FLAC^{3D} model. The figure clearly indicates the results of displacement, which were found

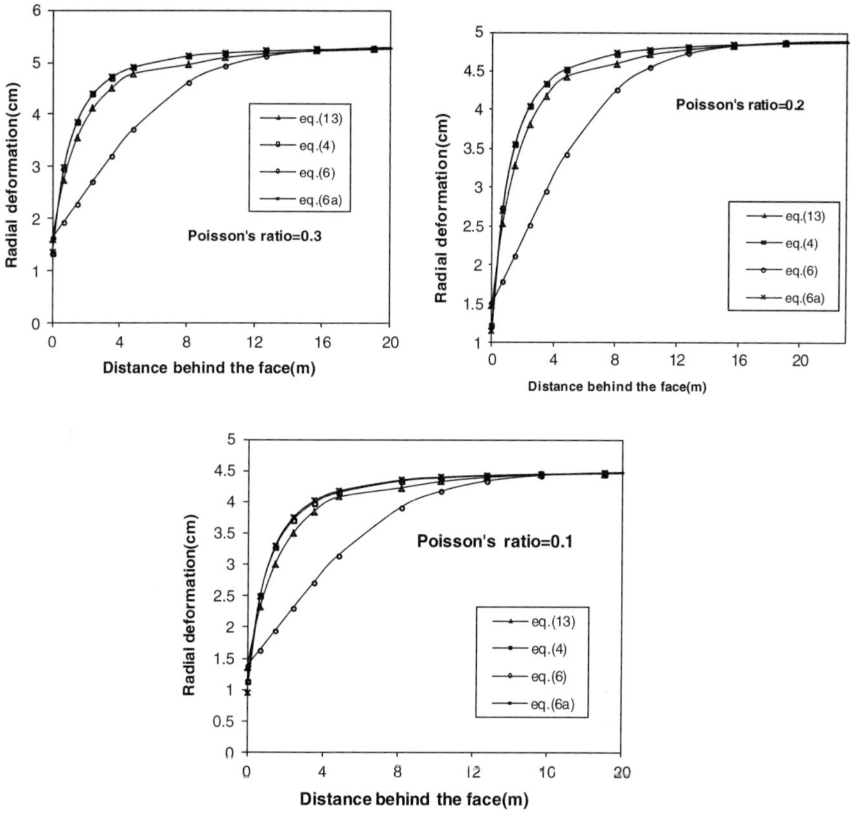

Fig. 13.11 (**a**–**c**) Comparison of empirical equations by Panet (1995), Fairhurst (2000), Unlu and Gercek (2003) and equation proposed by numerical $FLAC^{3D}$ model

numerically from the simulation and comparison from other researchers. The displacement is continuously increasing behind the face for region near the face. At the grid points ahead of face, there is drastic decrease comparatively from the displacement magnitude at face. After that, it decreases gradually and finally becomes insignificant. It can also be observed that, direction of material flow is in radial direction and tending towards vertical as the point of observation goes away from tunnel into the rock mass. The contour of radial displacement is shown in Fig. 13.13. It can be observed that, area of disturbance is elliptical around the tunnel rock mass. It also shows the region ahead of the face where there is practically no disturbance (no radial displacement). For this, a FISH function was incorporated in the model to plot contour of radial deformation which is resultant of displacement along x and z direction.

Figure 13.14 depicts the plot contour of volumetric strain increment around the tunnel face area. Volumetric strain signifies the increase or decrease of rock mass

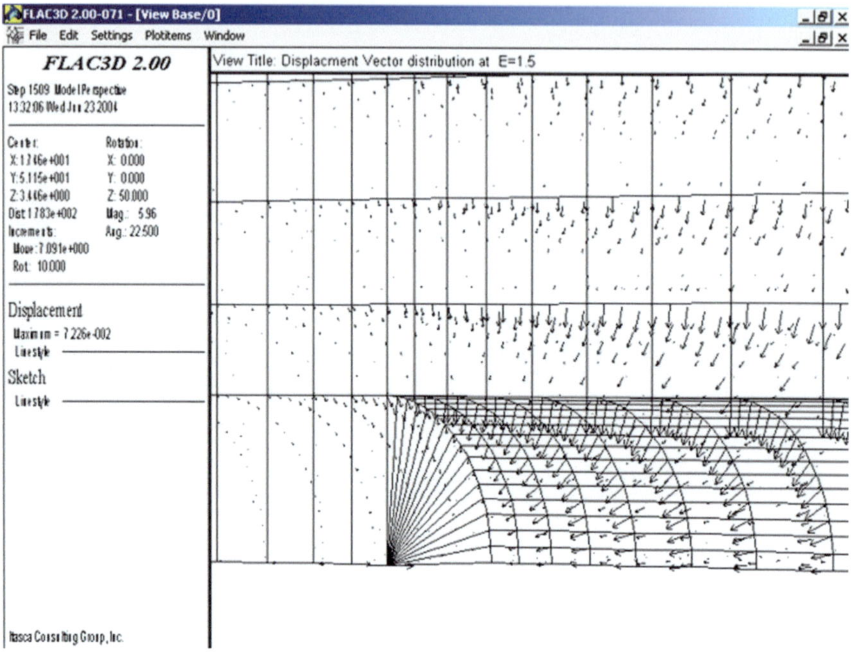

Fig. 13.12 Total displacement vector distribution around the tunnel face

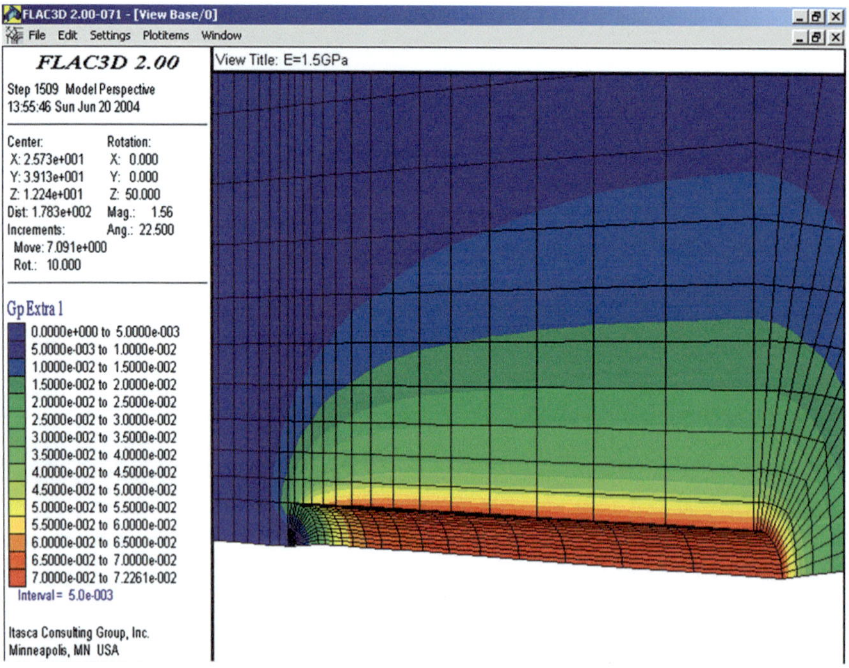

Fig. 13.13 Radial displacement in and around the circulated tunnel

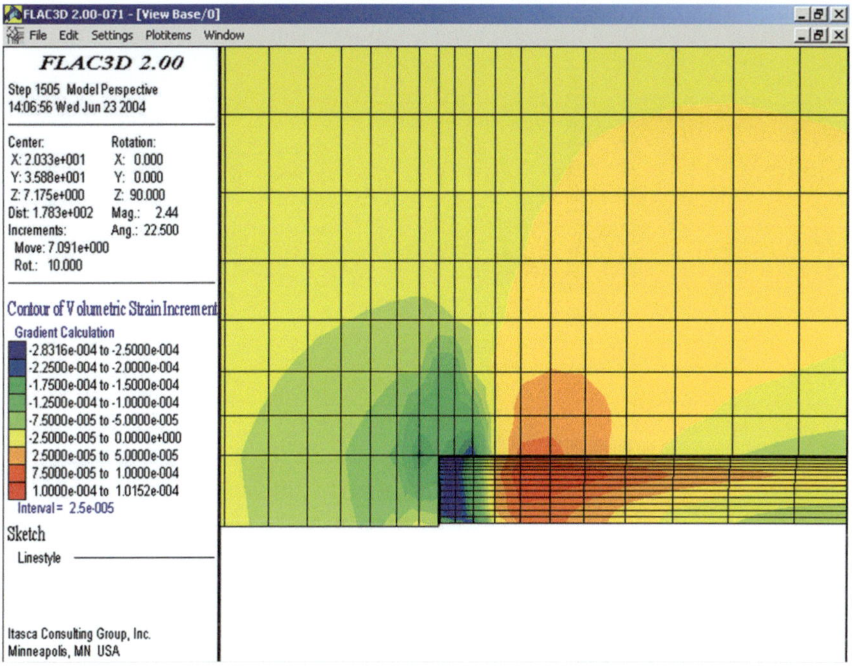

Fig. 13.14 Contour plot of volumetric strain distribution near the tunnel face

volume by opening or closing of pores in the rock mass. It can be observed that largest volumetric strains are developed near the tunnel face and decrease behind the face. It may be due to reason that, near the tunnel face, stresses developed are higher in magnitude then at other places. Similar observations are found for the region ahead of tunnel face. But magnitude of volumetric strain ahead of face is less in comparison to behind the face.

The variation of radial displacement along the whole profile of tunnel i.e. longitudinal deformational profile (LDP) is shown in Fig. 13.15. It can again be observed that there is a drastic decrease in deformation ahead of face. This suggests an increase in stability ahead of face in the rock mass. The displacement diminishes to almost zero ahead of face depending upon elastic properties as discussed earlier.

13.5 Conclusions

The present study reveals the stress/strain distribution pattern for a circular tunnel. The numerical observation were verified and tested with the available empirical relation proposed by various researchers. It has been observed that the radial deformation is inversely proportional to Young's modulus whereas linearly proportional to Poisson's ratio. Deformations in the rock mass near the tunnel face vanish

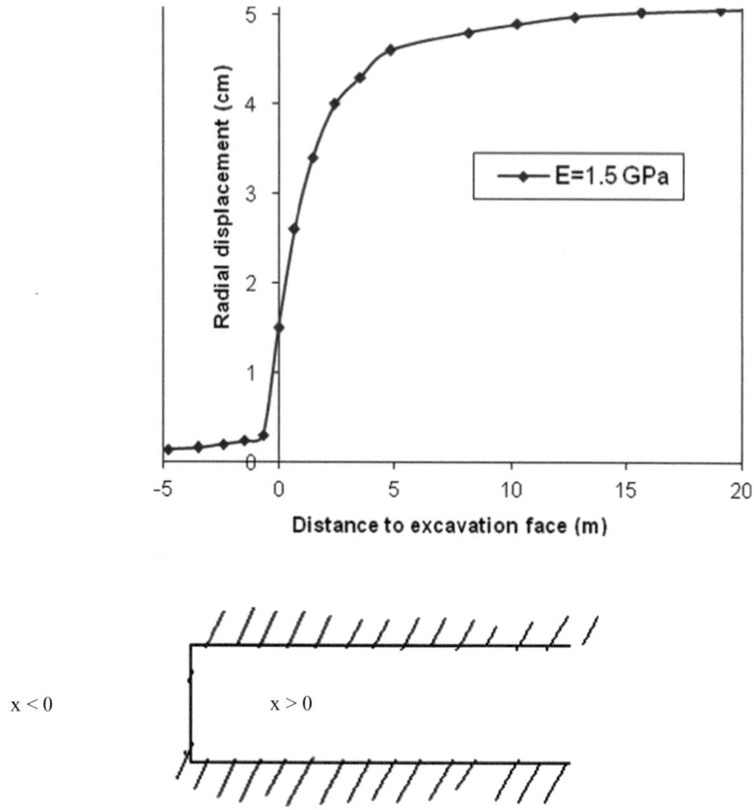

Fig. 13.15 Longitudinal deformational profile (*LDP*) or radial displacement

rapidly ahead of the face. On the other hand, it increases behind the face initially and later becomes constant until it reaches a critical distance. There is a close congruence between the analytical and numerical modeling results.

Also, the variation of radial deformation is found to be dependent on distance along the longitudinal axis of tunnel through sigmoid function at different Young's modulii and Poisson's ratio Critical distance is found to logarithmically depend on Young's modulus but independent of the Poisson's ratio. There is a continuous decrease in volumetric strain ahead and behind the tunnel face and it is maximum at the tunnel face.

References

Abel JF, Lee FT (1973) Stress changes ahead of an advancing tunnel. Int J Rock Mech Min Sci Geomech Abstr 10:673–679

Bieniawski ZT (1978) Determining rock mass deformability: experience from case histories. Int J Rock Mech Min Sci Geomech Abstr 15:237–247

Brady BHG, Brown ET (1985) Rock mechanics for underground mining. Allen & Unwin, London

Carranza-Torres C, Fairhurst C (2000) Application of the convergence confinement method of tunnel design to rock masses that satisfy the Hoek–Brown failure criterion. Tunn Undergr Space Technol 15(2):187–213

Corbetta F, Bernaud D, Nguyen-Minh D (1991) Contribution a la ' method convergerce–confinement par le principe de la similitude. Rev Fr Geotech 54:5–11

Guilloux A, Bretelle S, Bienvenue F (1996) Prise en compte des pre-soutenements dans le dimensionnement des tunnels. Rev Fr Geotech 76:3–16

Hoek E, Brown ET (1980) Underground excavations in rock. Institute of Mining and Metallurgy, London, pp 84–96

Itasca (1997) FLAC3D fast Lagrangian analysis of continua in 3 dimensions V.2.0. Itasca Consulting Group, Inc., Minnepolis

Jaeger JC, Cook NGW (1976) Fundamentals of rock mechanics, 2nd edn. Chapman & Hall, London

Niwa Y, Kobayashi S, Fukui T (1979) Stresses and displacements around an advancing face of a tunnel. In: Proceedings of the 4th international congress on rock mechanics, vol 1, ISRM. Balkema, Rotterdam, pp 703–710

Obert L, Duvall WI (1967) Rock mechanics and the design of structures in rock. Wiley, New York, 650 p

Panet M (1993) Understanding deformations in tunnels. In: Hudson JA, Brown ET, Fairhurst C, Hoek E (eds) Comprehensive rock engineering, vol 1. Pergamon Press, Oxford, pp 663–690

Panet M (1995) Calcul desT unnels par la Methode de Convergence–Confinement. Presses de l'Ecole Nationale des Ponts et Chaussees, Paris, p 178

Panet M, Guenot A (1982) Analysis of convergence behind the face of a tunnel. In: Proceedings of the international symposium on tunnelling, London, pp 197–204

Pellet F, Verma AK (2006) The effect of excavation rate on tunnel stability. Technical session 3H, A168. In: Second international congress on computational mechanics and simulation (ICCMS 06). IIT, Guwahati

Pelli F, Kaiser PK, Morgenstern NR (1991) The influence of near face behaviour on monitoring of deep tunnels. Can Geotech J 28:226–238

Sakurai S (1978) Approximate time-dependent analysis of tunnel support structure considering progress of tunnel face. Int J Numer Anal Methods Geomech 2:159–175

Serafim JL, Pereira JP (1983) Consideration of the geomechanics classification of Bieniawski. In: Proceedings of international symposium on engineering geology and underground constructions, Lisbon, pp 1133–1144

Singh TN, Verma AK (2007) Evaluating the slope instability of the Amiya slide. In: Eberhardt E, Stead D, Morrison T (eds) Rock mechanics. Meeting society challenges and demand. Taylor & Francis, London, pp 993–998

Singh TN, Verma AK (2009) Stress response around fracture surface under uniaxial loading. e-Journal Earth Sci India 2(3):148–159

Sitharam TG, Maji VB, Verma AK (2005) Equivalent continuum analyses of jointed rockmass. In: 40th U.S. symposium on rock mechanics (USRMS): rock mechanics for energy, mineral and infrastructure development in the northern regions, Alaska, pp 25–29

Sulem J, Panet M, Guenot A (1987) An analytical solution for time dependent displacements in a circular tunnel. Int J Rock Mech Min Geomech Abstr 24(3):155–164

Unlu T, Gercek H (2003) Effect of Poisson's ratio on the normalized radial displacement occurring around the face of a circular tunnel. Tunn Undergr Sp Technol 18:547–553

Verma AK, Singh TN (2010) Assessment of tunnel instability – a numerical approach. Int J Arab Geosci, Springer publication 3(2):181–192

Section IV
Natural Resource Management

Chapter 14
Gas Hydrate: A Viable Future Major Energy Resource of India

Kalachand Sain

14.1 Introduction

Depletion of fossil fuels and escalating demand of energy impose to search for an alternate source of energy for sustainable growth and development. Gas hydrate is a crystalline form of methane and water (Fig. 14.1a), and is a feasible resource of future due to its huge potential i.e., more than twice the energy content of total fossil fuels. Gas hydrate forms in shallow sediments of outer continental margins and the permafrost regions at high pressure and low temperature when methane concentration exceeds the solubility limit. One volume of gas hydrate releases about 164 volume of methane and 0.8 volume of fresh water (Fig. 14.1b) at standard temperature and pressure (STP). Presence of gas hydrate makes the sediments impermeable and hence the hydrate-bearing sediments trap 'free-gas' underneath. Unlike natural gas, oil and minerals, gas hydrate is not stable at STP.

The Gas-hydrate exploration in Indian offshore was first initiated by the Gas Authority of India Limited in 1996 under the aegis of the Ministry of Petroleum & Natural Gas. Various Institutes and industries joined the National Gas Hydrate Program (NGHP). The parameters such as the bathymetry, seafloor temperature, total organic carbon (TOC) content, sedimentary thickness, rate of sedimentation, geothermal gradient indicate good prospects of gas hydrate along the Indian continental margin (Ramana et al. 2006; Sain and Gupta 2008; Ramana et al. 2009). As a constituent member we, at the National Geophysical Research Institute (NGRI), Hyderabad have scrutinized more than 50,000 line km of single channel seismic records (analog), and detected probable sites for gas hydrate along the Indian margin(Gupta et al. 1998).

A total volume of 1894 trillion cubic meter of methane has been predicted in the form of gas hydrate within the Indian exclusive economic zone (EEZ). This

K. Sain (✉)
CSIR-National Geophysical Research Institute, Uppal Road, Hyderabad 500 007, India
e-mail: kalachandsain@yahoo.com

R. Sinha and R. Ravindra (eds.), *Earth System Processes and Disaster Management*, Society of Earth Scientists Series 1, DOI 10.1007/978-3-642-28845-6_14,

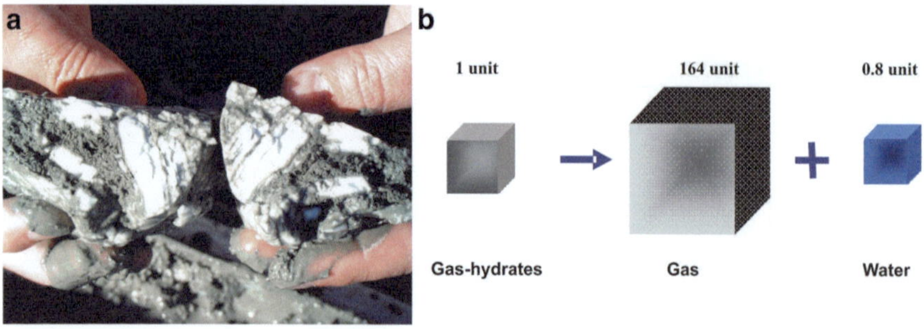

Fig. 14.1 (**a**) Gas-hydrate-bearing sediments, (**b**) one unit of gas-hydrates equivalents to 164 units of gas at STP

predicted volume is more than 1,500 times the country's current natural gas reserve. Even 10% recovery of methane from gas hydrate can meet the country's energy requirement for a century or so. Methane is the cleanest among all hydrocarbon fuels, hence can cause less pollution to the atmosphere. Therefore identification of gas hydrate deposits and evaluating their resource potential are of paramount significance. Earth Sciences in general and geophysics in particular, perhaps can provide much needed impetus for the exploration of this new energy resource. Most of the India's deep water regions remained unexplored. Therefore, the Ministry of Earth Sciences (MoES) launched a comprehensive research-oriented gas hydrate program for exploration of gas hydrate reserves using science and technology. Two CSIR institutions (NGRI and NIO) are involved in understanding the science component of gas hydrate reserves, while the NIOT (MoES) is involved in technology development for validation followed by exploitation of the gas hydrate deposits.

Interpretation of high resolution sparker data and geological samples in the Krishna-Godavari (KG) and Mahanadi basins show good signatures of gas hydrate (Dewangan et al. 2010; Mazumdar et al. 2009). To strengthen the understanding of spatial distribution of subsurface gas hydrate occurrence, NGRI acquired a large volume of multi-channel seismic (MCS) and ocean bottom seismic (OBS) data sets in the deep water (500–2,500 m) regions in these basins under the auspices of the MoES. The special processing and modelling of these data sets are in progress.

14.2 Identification of Gas Hydrate

The most commonly used acoustic marker for the detection of gas hydrate is an anomalous reflector, known as the bottom simulating reflector or BSR, which can be identified on seismic section. Figure 14.2 is one such example where BSR is detected based on its characteristics like mimicking the seafloor topography, exhibiting opposite polarity with respect to the seafloor event and crosscutting of the dipping reflections. In fact, the BSR is a physical boundary between the hydrate-bearing sediment above and gas-bearing sediments below. Using the

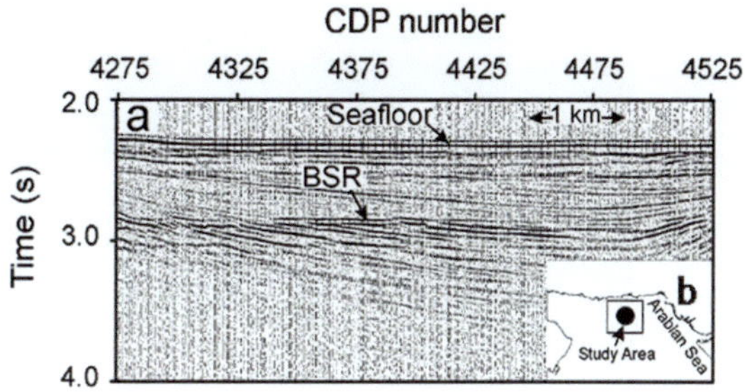

Fig. 14.2 A specimen seismic section showing BSR in the study area (Sain et al. 2000)

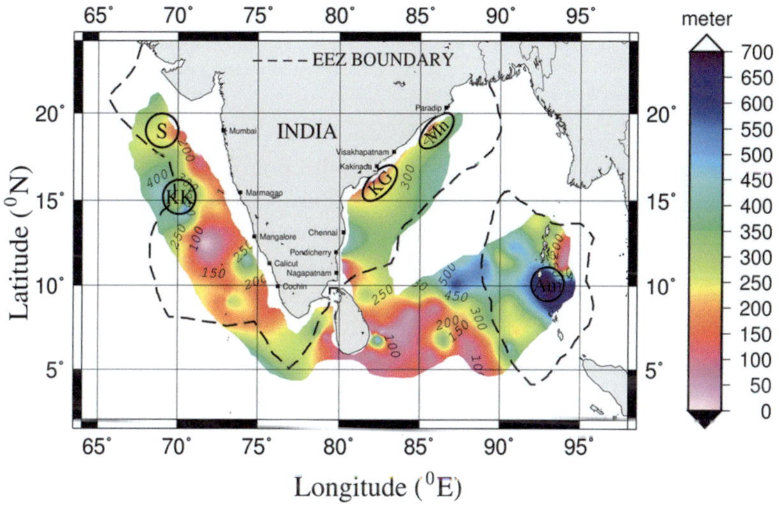

Fig. 14.3 Gas-hydrate stability thickness map along the Indian continental margin with potential zones marked by *circles/ovals* (Sain et al. 2011). *S* Saurashtra, *KK* Kerala-Konkan, *KG* Krishna-Godavari, *Mn* Mahanadi, *Am* Andaman regions

seafloor temperature, bathymetry and geothermal gradient data available till date, a modified gas hydrate stability thickness map (Sain et al. 2011) of India has recently been generated. Since the BSR is often associated with the base of gas hydrate stability zone, this map (Fig. 14.3) helps in detecting BSRs on seismic sections, and earlier map (Rao et al. 1998) facilitated in identifying gas hydrate in the KG, Mahanadi, Andaman Sea, Kerala-Konkan (KK) and Saurashtra regions (Sain and Gupta 2008).

The preliminary processing of newly acquired MCS data using the state-of-the-art pre-stack depth migration exhibits wide-spread occurrences of BSRs (Sain and Gupta 2012; Sain et al. 2012) thereby confirming the gas hydrate potential of the KG and Mahanadi basins. The drilling and coring under the NGHP Expedition-01 (Collett et al. 2008) confirmed the presence of gas hydrates in the KG, Mahanadi and Andaman basins. An overview of various approaches with field examples are available in several publications (Sain and Gupta 2008; Sain and Ojha 2008a).

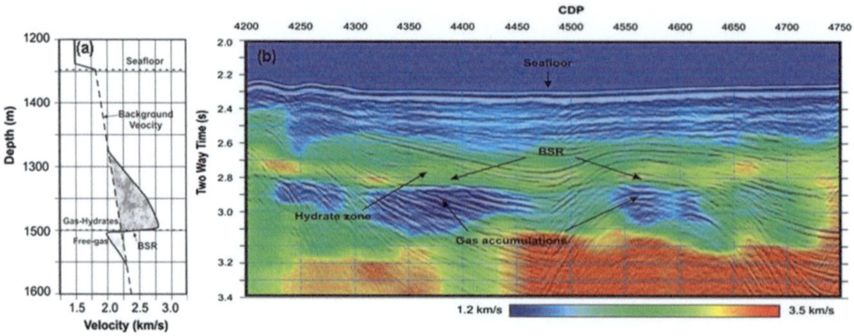

Fig. 14.4 (**a**) Typical 'velocity increase' of hydrates- and 'velocity reduction' of gas-bearing sediments against the background trend across the BSR, (**b**) seismic velocity anomaly showing the lateral and *vertical* distribution of gas-hydrates and free-gas bearing sediments (Ojha and Sain 2009) along the *seismic line* shown in Fig. 14.2

Seismic attributes such as the blanking, reflection strength and instantaneous frequencies (Satyavani et al. 2008; Ojha and Sain 2009) have been used to characterize the gas hydrate and free-gas bearing sediments. The presence of gas hydrate stiffens the sediment matrix thereby the hydrate bearing sediments exhibit low attenuation (Q^{-1}). Whereas, underlying gas-bearing sediments are associated with high attenuation. We have estimated the seismic attenuation and shown that gas-hydrate-bearing sediments are associated with low attenuation (Sain et al. 2009; Sain and Singh 2011), and thus can be used as a proxy for identifying gas hydrate. As the presence of gas hydrate increases and underlying free-gas decreases the seismic velocity with respect to the velocity of host sediments (Fig. 14.4a), the velocity anomaly (Fig. 14.4b) has been used for the delineation of gas hydrate and free gas bearing sediments across a BSR in the Makran accretionary prism. The travel time inversion of large-offset MCS data exhibits the lateral and vertical extension of gas hydrate and gas-bearing sediments in the KK basin (Singh and Sain 2012). Likewise, if we derive velocity anomaly from 3-D seismic data, we can demarcate the areal and vertical extension of gas-hydrate- and free-gas bearing sediments in an area.

We have shown that the pockmarks at seafloor or gas escape features in shallow sediments such as faulting or gas-chimney (Shankar and Sain 2007) can offer indirect evidences for gas hydrate. We have also brought out specific character of BSR near mud diapir and geophysical proxies (Satyavani et al. 2003; Shankar et al. 2008) for the investigation of gas hydrate in the western Indian margin.

14.3 Quantification of Gas Hydrate

We have estimated the velocity anomaly (Ojha and Sain 2009) across the BSR, AVO modeling (Ojha and Sain 2007) and full-waveform inversion (Sain et al. 2000), and quantified the saturations of gas hydrate and free-gas by

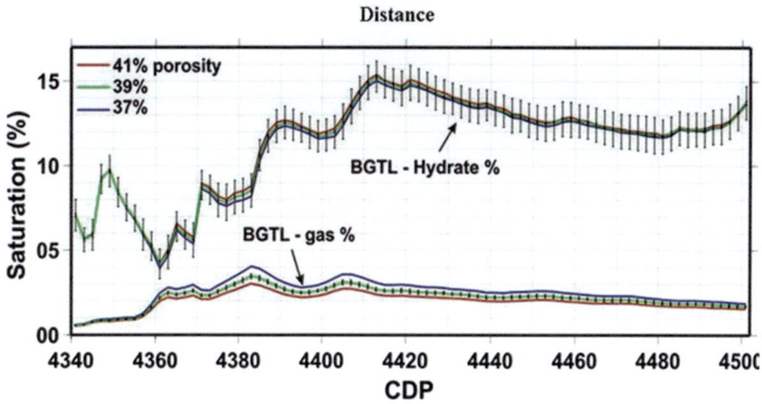

Fig. 14.5 Assessment of gas-hydrates and free-gas (Ojha et al. 2010) at BSR along the *seismic line* shown in Fig. 14.2

employing the rock physics modeling (Ghosh and Sain 2008; Ghosh et al. 2010a, b; Ojha and Sain 2008; Ojha et al. 2010; Sain et al. 2010) to the velocity anomaly. To assess the estimation of gas hydrate and free-gas, we have employed two different techniques to the same data set, and appraised 15.5% gas hydrate and 4.5% free-gas from cooperative travel time inversion followed by AVO modeling, and 13% gas hydrate and 2.8% free-gas from AVO attributes (Sain and Ojha 2008b). Using the AVO A-B crossplot coupled with the Biot-Gassmann Theory modified by Lee, we have assessed the saturations gas hydrate and free-gas varying between 4.5–15%, and 2.0–3.5%, respectively at BSR (Fig. 14.5) along a seismic line.

By employing the effective medium modeling to the fractured shale in the KG basin, for the first time we have estimated from sonic velocity the saturations of gas hydrate as 35–42%, 27–30% and 33–41% of total porosity respectively in the depth range between 60 and 140 m below sea floor for three different morphologies of pore-filling, grain displacing, and a combination of grain displacing and pore-filling respectively (Ghosh et al. 2010a). We have also estimated average saturations of gas hydrate as 10.5%, 11.5% and 12.5% of pore space respectively at sites 994, 995 and 997 in the clay-dominated Blake ridge area using the effective medium modeling of VSP data (Ghosh et al. 2010b). Earlier, we employed the technique in the sand-dominated Makran offshore (Ghosh and Sain 2008). The effective medium modeling can thus be employed to both the clay- and sand-dominated sediments. We have demonstrated with a case study that estimation of gas hydrate and free-gas using various rock physics model depends on mineralogy, porosity, and choice of background etc. (Sain et al. 2010). Using the constrained AVA modelling, we have quantified 30% gas hydrate in KK basin of the western Indian margin (Ojha and Sain 2007).

14.4 Concluding Remarks

We have proposed several approaches for the identification and quantification of gas hydrate, which is a prerequisite for further exploitation of gas hydrate. Analysis of the available and new data sets shows widespread occurrences of BSR in the KG, Mahanadi and Andaman regions in the Bay of Bengal, the ground truth of which has been validated by NGHP drilling. The Saurashtra and KK regions in the Arabian Sea also show indications for gas hydrate. We are in the process of evaluating the resource potential of gas hydrate in the prospective zones. For this purpose, we have acquired new data sets with suitable parameters in the KG and Mahanadi basins. We also plan for acquiring MCS and OBS data in the Andaman, Kerala-Laccadive and Cauvery regions. Not only for estimating the resource potential, the seismic data characterizes the sediments also, which provide inputs for the development of environment-friendly production technology. As the energy potential of gas hydrate is huge, this provides great hopes to overcome the present crisis of oil/gas. The Earth Scientists can play important role in this endeavour. If exploited properly, gas hydrate can meet the overwhelming demand of energy at one hand and reduce the environmental and submarine geo-hazards on the other hand. A concerted effort is most desirable to study the gas hydrate spanning from basic science to technology development in a systematic way to drive a country towards self sufficiency in the energy sector and a world leader in the field of gas hydrate exploration and production.

Acknowledgement I am grateful to the Director, NGRI for his permission to publish this work. The Ministry of Earth Sciences is acknowledged for financial support. I thank the staff at gas hydrate group for their contribution to the project.

References

Collett TS, Riedel M, Cochran J, Boswell R, Presley J, Kumar P, Sathe AV, Sethi AK, Lall M, Sibal VK, NGHP Expedition 01 Scientists (2008) NGHP Expedition 01 (2006). Initial Reports, Directorate General of Hydrocarbons, Noida and Ministry of Petroleum & Natural Gas, India, 4 volumes

Dewangan P, Ramprasad T, Ramana MV, Mazumdar A, Desa M, Badesab FK (2010) Seabed morphology and gas venting features in the continental slope region of Krishna-Godavari basin, Bay of Bengal: implications in gas–hydrate exploration. Mar Petrol Geol 27:1628–1641

Ghosh R, Sain K (2008) Effective medium modeling to assess gas hydrate and free gas evident from the velocity structure in the Makranaccretionary prism. Mar Geophys Res 29:267–274

Ghosh R, Sain K, Ojha M (2010a) Effective medium modeling of gas hydrate-filled fractures using sonic log in the Krishna-Godavari basin, eastern Indian offshore. J Geophys Res 115:1–15, B06101

Ghosh R, Sain K, Ojha M (2010b) Estimating the amount of gas hydrate using effective medium theory: a case study in the Blake ridge. Mar Geophys Res 31:29–37

Gupta HK et al (1998) Analysis of single channel seismic data from the continental margins of India for exploration of gas hydrates. NGRI Technical Report No. NGRI-98-LITHOS-221

Mazumdar A, Dewangan P, Joao HM, Peketi A, Khosla VR, Kocherla M, Badesab FK, Joshi RK, Roxanne P, Ramamurty PB, Karisiddaiah SM, Patil DJ, Dayal AM, Ramprasad T,

Hawkesworth CJ, Avanzinelli R (2009) Evidence of paleo–cold seep activity from the Bay of Bengal, offshore India. Geochem Geophys Geosyst 10:1–15
Ojha M, Sain K (2007) Seismic velocities and quantification of gas hydrates from AVA modeling in the western continental margin of India. Mar Geophys Res 28:101–107
Ojha M, Sain K (2008) Appraisal of gas hydrates/free-gas from V_P/V_S ratio in the Makranaccretionary prism. Mar Petrol Geol 25:637–644
Ojha M, Sain K (2009) Seismic attributes for identifying gas hydrates and free-gas zones: application to the Makranaccretionary prism. Episodes 32:264–270
Ojha M, Sain K, Minshull TA (2010) Assessment of gas hydrates saturation in the Makranaccretionary prism using the offset dependence of seismic amplitudes. Geophysics 75(2):C1–C6
Ramana MV, Ramprasad T, Desa M, Sathe AV, Sethi AK (2006) Gas hydrate-related proxies inferred from multidisciplinary investigations in the Indian offshore areas. Curr Sci 91:183–189
Ramana MV, Ramprasad T, Paropkari AL, Borole DV, Rao BR, Karisiddaiah SM, Desa M, Kocherla M, Joao HM, Lokabharati P, Gonsalves MJ, Pattan JN, Khadge NH, Babu CP, Sathe AV, Kumar P, Sethi AK (2009) Multidisciplinary investigations exploring indicators of gas hydrates occurrence in the Krishna–Godavari basin offshore, east coast of India. Geo-Mar Lett 29:25–38
Rao HY, Reddy SI, Khanna R, Rao TG, Thakur NK, Subrahmanyam C (1998) Potential distribution of methane hydrates along the Indian continental margins. Curr Sci 74:466–468
Sain K, Gupta HK (2008) Gas hydrates: Indian scenario. J Geol Soc India 72:299–311
Sain K, Ojha M (2008a) Identification and quantification of gas hydrates: a viable source of energy in the 21st century. Mem Geol Soc India 68:273–288
Sain K, Ojha M (2008b) Estimation of gas hydrates and free-gas concentrations using modeling and crossplot of seismic amplitudes from the bottom simulating reflector. Adv Geosci Ocean Sci 18:181–196
Sain K, Singh AK (2011) Seismic quality factors across a bottom simulating reflector in the Makranaccretionary prism, Arabian Sea. Mar Petrol Geol 28:1838–1843
Sain K, Gupta HK (2012) Gas hydrates in India: Potential and Development. Gond Res, in press, doi:10.1016/j.gr.2012.01.007
Sain K, Minshull TA, Singh SC, Hobbs RW (2000) Evidence for a thick free-gas layer beneath the bottom-simulating reflector in the Makranaccretionary prism. Mar Geol 164:3–12
Sain K, Singh AK, Thakur NK, Khanna RK (2009) Seismic quality factor observations for gas-hydrate-bearing sediments on western margins of India. Mar Geophys Res 30:137–145
Sain K, Ghosh R, Ojha M (2010) Rock physics modeling for assessing gas hydrate and free gas: a case study in the Cascadia accretionary prism. Mar Geophys Res 31:109–119
Sain K, Rajesh V, Satyavani N, Subbarao KV (2011) Gas hydrates stability thickness map along the Indian continental margin. Mar Petrol Geol 28:1779–1786
Sain K, Ojha M, Satyavani N, Ramadass GA, Ramprasad T, Das SK, Gupta HK (2012) Gas-hydrates in Krishna-Godavari and Mahanadi basins: new data. J Geol Soc India 79: 553–556
Satyavani N, Thakur NK, Shankar U, Reddi SI, Sridhar AR, Rao PP, Sain K, Khanna R (2003) Indicators of gas hydrates: role of velocity and amplitude. Curr Sci 85:1360–1363
Satyavani N, Sain K, Malcolm Lall, Kumar BJP (2008) Seismic attribute study for gas hydrates in the Andaman offshore, India. Mar Geophys Res 29:167–175
Shankar U, Sain K (2007) Specific character of the bottom simulating reflectors near mud diapirs: western margin of India. Curr Sci 93:997–1002
Shankar U, Ojha M, Sain K, Khanna RK, Sudhakar M, Tyagi A (2008) Seafloor geophysical study in search of gas hydrates/gas related evidences in the deep waters of the western continental margin of India. J Geol Soc India 72:547–555
Singh PK, Sain K (2012) 2-D Velocity structure in Kerala-Konkan basin using travel time inversion of seismic data. J Geol Soc India 79:53–60

Chapter 15
Impact Cratering from an Indian Perspective

Jayanta K. Pati and Puniti Pati

15.1 Introduction

Bolide impact structures broadly cover the surfaces of planetary bodies in the solar system (Taylor 1992) and occur in the planets of the inner solar system, their moons, the icy satellites and moons of the of the outer solar system planets, comets and asteroids. Nearly 70% of the Moon's surface is covered by impact structures. However, these are rare features on the surface of the planet Earth and only 176 structures are currently known (http://www.unb.ca/passc/ImpactDatabase/Age; April 25, 2011) as confirmed impact structures. The discovery of meteoritic impact structures dates back to 1905 and a sharp rise in their numbers took place in 1960 with the increase in the awareness of various diagnostic petrographic and mesoscopic shock metamorphic features (French and Koeberl 2010). Presently, the rate of discovery is about 3–4 structures per year (Grieve and Shoemaker 1994). Unfortunately, the newly reported structures are relatively young and only 17 structures of older than 600 Ma are known (http://www.passc.net/EarthImpactDatabase/Agesort.html; 25:04:2011). Most impact structures known on Earth are younger than 300 million years and smaller than ~20 km in diameter (Grieve et al. 1995; Reimold and Gibson 1996). Only a few much larger and older structures have been recognized on Earth. This includes the 2,025 $\pm$ 4 Ma old, originally 280 km wide, Vredefort Structure in South Africa (Kamo et al. 1996; Stöffler et al. 1994) and the ~1,850 Ma old Sudbury Structure in Canada (about 200 km wide; Deutsch et al. 1995).

The recognition of relatively young and well-preserved impact structures may be comparatively easy, provided diagnostic evidences. The identification of old deeply eroded impact structures is often difficult (Grieve et al. 1996; Koeberl et al. 2004). In case of a perfectly circular, regionally distinct structure, remote sensing and

J.K. Pati (✉) • P. Pati
Department of Earth and Planetary Sciences, Nehru Science Centre, University of Allahabad, Allahabad 211002, India
e-mail: jkpati@yahoo.co.in

R. Sinha and R. Ravindra (eds.), *Earth System Processes and Disaster Management*, Society of Earth Scientists Series 1, DOI 10.1007/978-3-642-28845-6_15,

Fig. 15.1 The distribution of Asian impact structures including those in Russia (http://www.unb.ca/passc/ImpactDatabase/Age; April 25, 2011). Recently discovered Mt. Oikeyama crater, Japan (Sakamoto et al. 2010) has been added in the image by the author

geophysical techniques may help in the initial detection of possible impact structures (Pilkington and Grieve 1992; Donofrio 1997; Grieve 2005; Koeberl and Reimold 2005; Koeberl et al. 2005; Buchner and Kenkmann 2008). However, the confirmation of an impact structure requires the presence of shock metamorphic features in meso- and microscopic scales, such as planar deformation features in the rock forming minerals, shatter cones, and/or fragments of the impacting projectile.

The bolide impacts generate enormous energy in terms of temperature (>20,000°C) and pressure (>100 GPa). They help in the generation and remobilization of the ore-bearing fluids and consequent mineralization. It is important to note that nearly 60% of the terrestrial impact structures are associated with some form of economic mineral deposits (Au, Ni-PGE, and U) or natural resources which include some of the important hydrocarbon deposits. The large Chicxulub impact structure, Mexico (~190 km in diameter), for instance, produces ~2.1 million barrels of oil per day (Reimold et al. 2005). However, even the small impact structure of View field, Saskatchewan, Canada (~2.5 km in diameter), produces significant amounts of oil and gas (~600 barrels of oil and ~250 million cubic feet of gas per day; Dietz 1947). In addition, the process of impact cratering is known to have possibly contributed to the presence of life and water on Earth although it is hard to establish this with the available evidence at hand.

Although the Asian impact record looks impressive, it is mainly due to structures present in the Russian mainland (Fig. 15.1; Table 15.1). The Lonar structure (Fredriksson et al. 1973; Murali et al. 1987; Nayak 1972, 1993; Osae et al. 2005) in Maharashtra and Dhala structure, Madhya Pradesh (Pati 2005; Pati and Reimold 2007; Pati et al. 2008a, 2010) are until now the only confirmed impact structures in the entire region of the Indian subcontinent. The Ramgarh structure, a crater-like feature in Rajasthan (Master and Pandit 1999; Sisodia et al. 2006), and the proposed

Fig. 15.3 The Tswaing crater (aka Pretoria Saltpan), Pretoria, South Africa, is a classic simple impact structure in a crystalline basement target

the shape and dimensions of the original transient cavity. The transient crater is modified only by minor collapse of the steep walls into the crater cavity, and by deposition of a minor amount of ejected material in the crater (Grieve et al. 1995). Complex craters (Fig. 15.4), on the other hand, possess central peaks, or an inner "peaked" ring, a flat floor, outer concentric faulted zones and terraced rim walls (Melosh 1989). It has been observed that with increasing diameter, the complex structures can also be delineated into following three types: central peak structures, central-peak-basin structures and peak-ring basin structures (Grieve et al. 1981; Melosh 1989; Spudis 1993). Multiring basins are the largest of impact structures (few hundred to >1,000 km; Schultz and Merrill 1981; Melosh 1989; Spudis 1993). However, there are air blasts devoid of any characteristic structures, for example Tunguska, Russia and no meteoritic material was recovered from Tunguska, and second, it (likely) did not cause a crater (Gasperini et al. 2007).

The third type is known as multiple ring basins and essentially comprises multiple concentric uplifted rings and intervening down-faulted valleys (ring grabens). This type of impact structure is invariably larger compared to other types of impact structures.

The confirmation of an impact structure is not easy but there are excellent papers available similar to a "Do-It-Yourself" manual (French 1998; French and Koeberl 2010 and references therein). Terrestrial impact structures are identified by the evidences such as:

1. Morphology of the crater (Non-diagnostic)
2. Geophysical anomalies (Non-diagnostic; especially magnetic and gravity anomalies)
3. Shock metamorphic features (Diagnostic), and
4. Physical presence of the bolide or its geochemical signature (Diagnostic).

Morphological characteristics provide only preliminary information for the identification of a possible impact structure. Geological structures with a circular

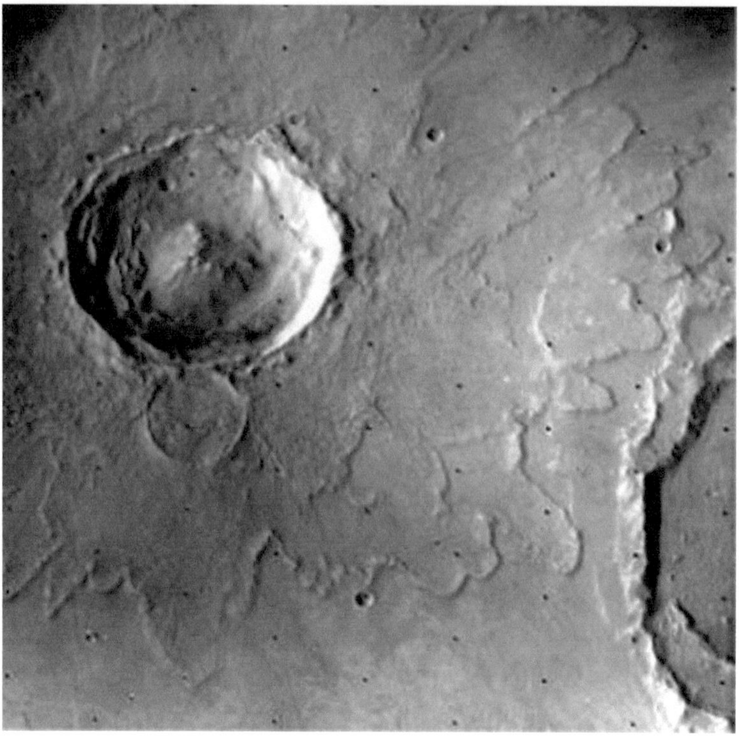

Fig. 15.4 A complex crater about 20 km diameter in the northern region of Mars. It has a large central peak and terraces around its rim. The ejecta blanket has lobes possibly suggesting that subsurface water or melted ice was mixed into the debris (Image from the Viking Orbiter, courtesy of the Lunar and Planetary Institute)

(near circular) outline at least deserve additional attention, especially areas where no other obvious mechanism for producing such features exists. Morphological observations can be better understood by remote sensing and geophysical studies in addition to field work, especially in case of deeply eroded impact structures, with better resolution and more spectral information data. Morphological and structural criteria can be applied to high-resolution images taken from space. Shock metamorphic evidences can be expressed in macroscopic form as shatter cones (Fig. 15.5) or in microscopic form as planar deformation features (PDFs), ballen quartz, checkerboard feldspar, high pressure mineral polymorphs, and diaplectic glasses. Shatter cones are the only megascopic (hand specimen to outcrop) feature that provide unequivocal evidence of shock metamorphism. They indicate shock pressures between ~5 and 15 GPa, and oftentimes occur in exposed central uplifts and their proper identification is necessary (French and Koeberl 2010). Low shock pressures (2–10 GPa) produce distinguishing conical fracturing patterns in the target rocks, and the resulting shatter cones have proven to be a reliable field criterion for identifying and studying impact structures (Dietz 1959, 1963, 1968; French and Short 1968; Milton et al. 1972). Shatter cones occur as individuals or

Fig. 15.5 The shatter cones seen in the photograph, Vredefort structure, South Africa (Photo: J.K. Pati) represent the only diagnostic megascopic shock feature

composite group cones and their lengths vary from millimeters to meters (Milton et al. 1996a). Some minerals of target rocks (e.g., quartz, graphite) may transform to high-pressure polymorphs due to shock metamorphism. Graphite (C) converts to diamond (cubo-octahedral) or lonsdaleite (hexagonal). However, the use of nanodiamonds as impact diagnostic criterion is still unclear (French and Koeberl 2010). Quartz is transformed to stishovite at shock pressures of >12–15 GPa and to coesite at >30 GPa (French and Short 1968). Shock waves create a variety of unusual features such as PDFs, ballen textured SiO_2 and partial to complete melting, etc. PDFs in quartz, feldspar, zircon and other minerals are the most subsequent identifications features of impact structures. These features characteristically occur as sets of parallel deformation planes filled with shock-produced glass within individual grains of the mineral. Distinctive planar features in quartz (SiO_2) have been one of the most widely applied criteria for recognizing impact structures (Stöffler and Langenhorst 1994; French and Koeberl 2010; Koeberl and Martinez-Ruiz 2003; Montanari and Koeberl 2000; French 1998; Grieve et al. 1996) (Fig. 15.6a). In contrast to planar fractures, with which they may occur, PDFs are not open cracks. Instead, they occur as multiple sets of closed, extremely narrow, parallel planar regions. Individual PDFs are both narrow (typically <2–3 μm) and closely spaced (typically 2–10 μm). In altered, geologically old, or metamorphosed samples, PDFs have an equally distinctive but discontinuous character. The original amorphous material in the PDF planes is recrystallized back to quartz, and in the process, arrays of small (typically 1–2 μm) fluid inclusions ("decorations") develop along the original planes (Fig. 15.6b). The resulting features are called decorated PDFs (French and Short 1968). The PDFs in zircon (Kamo et al. 1996) remains a contentious issue (Axel Wittmann, pers. Communication). The identification of PDF in zircons is not very well founded (e.g., Leroux et al. 1999) and although planar features occur in zircon at the microscopic and sub-microscopic level, they do not compare that well to quartz. Ballen quartz (Carstens 1975; Bischoff and Stöffler 1984; Ferrière et al. 2010) is an impact related microscopic features

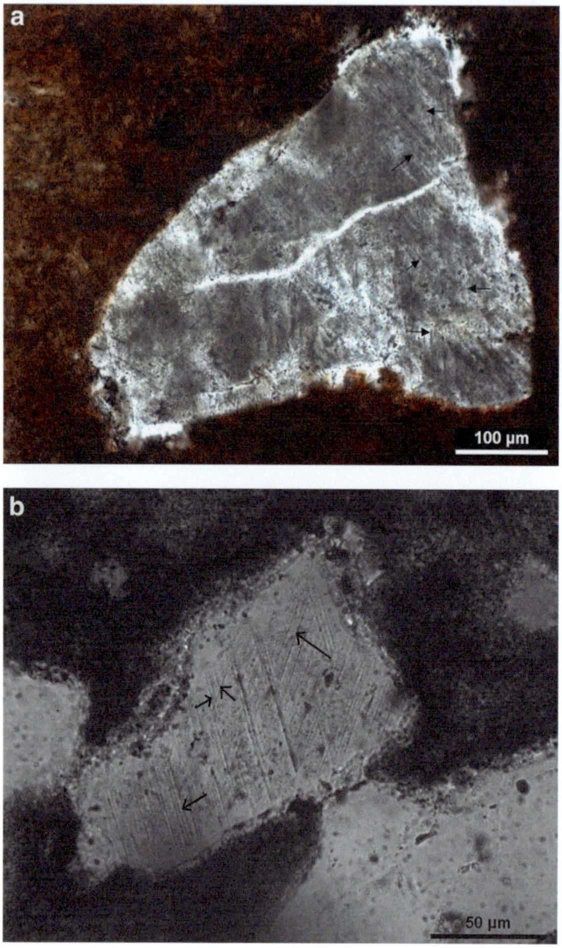

Fig. 15.6 (**a**) Microphotograph showing multiple sets of planar deformation features (PDFs; shown with *arrows*) in quartz grains in an aphanitic groundmass of impact melt breccia, Dhala structure, India. The brownish portions are described as "toasted" domains (Photo in plane polarized light: J.K. Pati). (**b**)The microphotograph shows two sets of decorated PDFs in quartz grains in an aphanitic groundmass (not seen in the photograph) of impact melt breccia, Dhala structure, India (Photo in plane polarized light: J.K. Pati).The PDFs are shown with *arrows*

consisting of a series of closed or open loops that range in diameter from 12 to 142 µm (Fig. 15.7) and considered as diagnostic of impact cratering process forming at pressures between 35 and 50 GPa with temperature in excess of 1,200–1,400°C. According to Bischoff and Stöffler (1984) the ballen represented recrystallized diaplectic quartz glass which had undergone the transition to cristobalite and then transformed back to α-quartz. Other, but equally definitive, microscopic indicators of impact include diaplectic glasses, in which feldspar is transformed to maskelynite. Lechatelierite (35 GPa), is an amorphous silica glass formed by shock-induced thermal transformation of quartz. In addition, the high-pressure mineral phases like coesite, stishovite, impact diamond and lonsdaleite

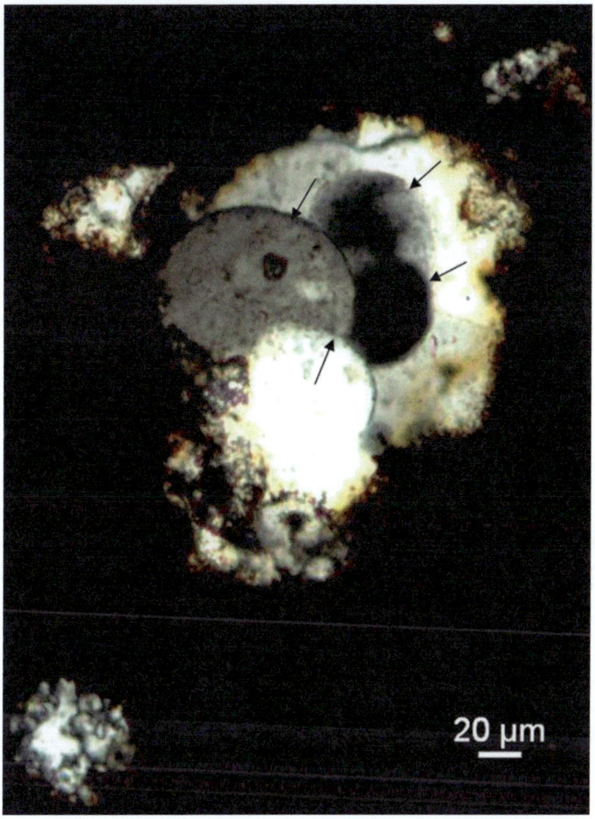

Fig. 15.7 Microphotograph under cross polars of a quartz grain occurring within impact melt breccia, Dhala structure, India showing open and closed loop-like structures (*ballen texture*). The ballen loops (shown with *arrows*) show varied optical orientations (Photo in cross polars: J.K. Pati)

occurring in impact melts and some meteorites are also considered impact diagnostic. Coesite, stishovite and impact diamond/lonsdaleite (18 GPa) are mostly found in rocks in shock stage II and III; in impact melted (shock stage IV) rocks, the high post-shock temperature regime typically causes reversion to the low-pressure polymorphs. In addition to the presence of megascopic and microscopic indications of shock metamorphism, geochemical analyses may provide definite evidence of impact by identifying a signature from the projectile, either excess iridium or distinctive osmium isotope ratios.

15.1.2 Impact Structures in India

The impact cratering research is still in infancy in India, even though we have two very important impact structures from the continental landmass of India. The lack

Fig. 15.8 Lonar structure, Buldhana district, Maharastra state, India is a rare simple crater on basaltic rock similar to Martian structures and is very well preserved (Photo: J.K. Pati)

of proper course/training pertaining to the identification and confirmation of impactites is one of the possible reasons. At the University of Allahabad, we have been successful to establish a good collection of impactites from many impact structures from different parts of the world and some of the basic equipments essential for impact cratering research with the help from premier government agencies like the Atomic Minerals Directorate for Exploration and Research (AMD), Department of Space (PLANEX programme) and Department of Science and Technology. In addition to the detailed study of Dhala structure, north central India, we are also involved in the study of Lonar structure, Maharastra and Ramgarh structure, Rajasthan.

15.1.3 Lonar Crater

It is centered at 19°58′N, 76°31′E in Buldhana district of Maharastra state, India (Fig. 15.8) with a rim-to-rim diameter of 1.88 m and nearly 135 m deep. The Lonar structure has a 20 m high rim surrounding a nearly 7 m deep salt water lake occupying the crater floor. It is an extremely well preserved, simple impact structure on basaltic rocks. The age of the Lonar structure is 564 ± 46 ka (Jourdan et al. 2010). It is a classic terrestrial analogue which provides unique opportunity to understand Martian and Lunar craters. Maloof et al. (2010) carried out geological mapping of Lonar Crater and the nearby Little Lonar (Amber Lake), structural deformation around the crater rim and studied the physical properties of the ejecta blanket. Kumar (2005) made a detailed analysis of structural deformation of the Lonar crater and showed listric faulting and slumping of the crater wall. The basaltic host rock is of tholeiitic affinity, showing porphyritic texture with phenocrysts of plagioclase and rare olivine that are set in a groundmass of plagioclase, augite, pigeonite, titanomagnetite and palagonite (Osae et al. 2005). The secondary minerals such as calcite, zeolite, chlorite, serpentine and chlorophaeite are also present (Osae et al. 2005). The evidences of impact cratering have mostly come from the presence of maskelynite, planar deformation features in plagioclase and pyroxene (Kieffer et al. 1976) and the other supporting evidences include the presence of impact melt glass, ejected melt breccias with shocked minerals, and a subsurface breccia observed in drill cores below the crater floor (Nayak 1972; Fredriksson et al. 1973, 1979; Fudali et al. 1980).

Fig. 15.9 The Dhala structure, Shivpuri district, Madhya Pradesh is a complex impact structure with a central elevated area in the background and a monomict breccia of granitoids is seen in the foreground. (Photo: J.K. Pati)

15.1.4 Dhala Structure

The Dhala impact structure is situated to the east of Dhala village, Shivpuri district, Madhya Pradesh state, India (Fig. 15.9). It is centered at N 25°17′59.7″ and E 78°8′3.1″ (Pati 2005; Pati and Reimold 2007; Pati et al. 2008a). Dhala is only the second confirmed impact structure in India, after the Lonar crater. It is a well preserved remnant of a large (estimated diameter of ~25 km; Pati et al. 2008b), complex impact structure, which occurred on Archean crystalline basement (Bundelkhand granitoid; 2,563 ± 6 Ma; Pati et al. 2010). The Dhala structure is partially covered, in its central portions, by post-impact sediments presumed to belong to the Vindhyan Supergroup (1,729 ± 110 Ma; Sarangi et al. 2004). More than 50 locations in and around the central elevated area have been drilled by various governmental agencies in the last 5 years.

A remote sensing study based on SRTM and IRS-1D (Indian Remote Sensing Satellite) data show a circular feature with an estimated diameter of about 25 km with an asymmetrically disposed central elevated area suggesting an oblique impact. It shows an annular drainage pattern despite extensive erosion. Structural data obtained on the ground show a radial fracture pattern and radial displacement of the pre-impact fabric. Anisotropic magnetic susceptibility (AMS) measurements of impact melt breccia suggest essentially a random fabric. Fractal analysis of clasts in melt breccia samples in meso-, micro- and sub-microscopic scales shows that the clast size distribution follows a modified power law. The target rocks are predominantly composed of different textural variants of granitoids including rhyolites with enclaves of older meta-supracrustals (calc-silicate rocks, quartzite, amphibolite). Impact melt breccia occurs above fractured granitoid basement with a sharp but undulatory contact. Clasts in the impact melt breccia contain abundant unequivocal evidence of shock metamorphism (Figs. 15.6a, b and 15.7), including completely molten rock clasts (Fig. 15.10). Recently mm to cm wide pseudotachylitic breccia veins, sometimes occurring in networks, have been observed at different depths in the fractured granitoid basement (Pati et al. 2009). The matrices of these veins are extensively altered. The shock deformation features are observed in some of these veinlets contain

Fig. 15.10 The photomicrograph in plane polarized light showing several mineral and rock clasts of angular to subangular shape in a melt matrix, Dhala structure, Shivpuri district, Madhya Pradesh. The large lithic clast exhibiting perlitic texture is characteristic of complete rock melting. The flow texture in the *dark brown* melt matrix is also very distinct (Photo in plane polarized light: J.K. Pati)

clasts with. The surrounding host rock is fractured. The contact between impact melt breccia and overlying post-impact sediments is a sharp, erosional contact.

The diagnostic evidences of shock metamorphism observed in Dhala impact melt breccia include one to five sets of PDFs in quartz (Fig. 15.6a, b) and feldspar clasts, besides ballen quartz (Fig. 15.7), checkerboard feldspar, and granular zircon. The presence of the high-pressure polymorph reidite ($ZrSiO_4$ in scheelite crystal structure) has been confirmed by Raman spectroscopy on impact melt rock samples. While Ir content (0.1–0.8 ppb) in a limited number of analyses of impact melt breccia shows a positive correlation with Ni/Cr ratios (0.24–1.99), most values are not elevated above typical crustal values. The major, minor and trace element (including REEs) compositions of host rock and impact melt breccia are similar.

15.1.5 Ramgarh Structure

The Ramgarh structure (25°20′8.9″; 76°36′23.7″) located in the Baran district of Rajasthan, India is known since 1869 (Mallet 1869). It can be seen prominently from a distance of tens of km (Fig. 15.11) and can be easily picked up on satellite images as well. It is a near circular structure with a rim-to-rim diameter of ~4 km and a rim height of ~250 m. The structure comprises sedimentary rocks (sandstone, shale and limestone) belonging to the Vindhyan Supergroup of Neoproterozoic age (Ramasamy 1987). These rocks show extensive brittle failure with variable dips (5–50°) showing a quaquaversal dipping pattern away from the center – unlike the

Fig. 15.11 The Ramgarh structure (~4 km diameter), Rajasthan can be clearly seen from a distance of about 5 km from the eastern side (Photo: J.K. Pati)

sub-horizontally oriented, largely undeformed Vindhyan Supergroup rocks exposed regionally. This structure has been drilled by the Geological Survey of India and uplifted shale and siltstone showing disharmonic folding and brecciation have been reported (Ramasamy 1987, 1995) earlier. The origin of this unusual structure has been attributed either to tectonic/volcanic activity (Ramasamy 1995; Sharma 1973) or to an impact event (Crawford 1972; Balasundaram and Dube 1973; Master and Pandit 1999; Sisodia et al. 2006). However, the unequivocal evidence(s) of shock metamorphism to confirm an impact origin is still lacking (Reimold et al. 2006). Recent study (unpublished data) of various rock types in and around Ramgarh structure by the author involving field and petrographic studies does not show the presence of PDFs reported earlier (Sisodia et al. 2006) although curved fratures, cataclastic veins and Boehm lamellae are in plenty. No diagnostic megascopic shock feature has been observed either.

15.1.6 Impact Cratering Research in India

India with an extensive, well exposed cratonic area of Archaean-Proterozoic age has so far been able to locate only two confirmed meteoritic impact structures. Recent studies have proven that Lonar crater is an excellent analogue for Mars and Dhala is a very well preserved Early Proterozoic crater formed on Archaean basement gneiss. It is necessary that the terrestrial crater search and impact cratering research in India should be encouraged and given priority so that a large pool of scientists are available to study available data pertaining to the search of new craters and evaluate the Lunar and Martian craters. It would be interesting to study all our cratonic areas to look for structures similar to Dhala. Radhakrishna (2005) has appealed to re-examine certain areas having circular to semi-circular morphology from impact cratering research viewpoint in different parts of India ((1) Cuddapah Basin, (2) the volcanic group of islands off the coast of Mumbai, and the islands of Saurashtra and Kachchh, (3) the regions of Malani and (4) the Simlipal Complex, Odisha). In addition to Dhala and Lonar, taking the examples of impact craters in Australia and Scandinavia, it appears that India could have a large potential for amending the terrestrial impact record. It seems that some of these old basement surfaces could have been saturated with craters and offers a window into the projectile flux in the Pre-Cambrian. The planetary exploration research (PLANEX) programme of the Department of Space, Government of India is funding impact cratering research in India. The large amount of data generated during our Chandrayaan-1 mission need scientific manpower trained in impact cratering research to draw meaningful scientific conclusions. Unfortunately, impact cratering research and simple petrographic techniques involved for the confirmation of meteoritic impact craters are not known in most universities and institutes of higher learning in India. Impact cratering research being a fundamental geoscience discipline, which has greatly influenced the scientific research in general, especially in the formation of planets, planetary systems and landscapes, requires immediate attention in India at all levels of education and research.

Acknowledgements JKP thanks the PLANEX, Department of Space, Government of India for funding the impact crater research at the University of Allahabad. We thank Dr. Axel Wittmann, LPI, USA for his excellent review and helpful suggestions which improved the overall quality of the manuscript.

References

Balasundaram MS, Dube A (1973) Ramgarh structure, India. Nature 242:40

Bischoff A, Stöffler D (1984) Chemical and structural changes induced by thermal annealing of shocked feldspar inclusions in impact melt rocks from Lappajärvi crater. Finland. J Geophys Res 89:B645–B656

Buchner E, Kenkmann T (2008) Upheaval dome, Utah: impact origin confirmed. Geology 36:227–230

Carstens H (1975) Thermal history of impact melt rocks in the Fennoscandian shield. Contrib Miner Petrol 50:145–155. doi:10.1007/BF00373334

Crawford AR (1972) Possible impact structure in India. Nature 237:98

Chatterjee S, Guven N, Yoshinobu A, Donofrio R (2006) Shiva structure: a possible KT boundary impact crater on the western shelf of India. Texas Technical University Natural Science Research Laboratory Special Publication # 50, 39pp

Deutsch A, Grieve RAF, Avermann M, Bischoff L, Brockmeyer P, Buhl D, Lakomy R, Müller-Mohr V, Ostermann M, Stöffler D (1995) The Sudbury structure (Ontario, Canada): a tectonically deformed multi-ring impact basin. Geol Rundsch 84:697–709

Dietz RS (1947) Meteorite impact suggested by the orientation of shatter-cones at the Kentland, Indiana disturbance. Science 105:42–43

Dietz RS (1959) Shatter cones in cryptoexplosion structures (meteorite impact). J Geol 67:496–505

Dietz RS (1963) Cryptoexplosion structures: a discussion. Am J Sci 261:650–664

Dietz RS (1968) Shatter cones in cryptoexplosion structures. In: French BM, Short NM (eds) Shock metamorphism of natural materials. Mono Book Corporation, Baltimore, pp 267–285

Donofrio RR (1997) Survey of hydrocarbon-producing impact structures in North America: exploration results to date and potential for discovery in Precambrian basement rock. Oklahoma. Geol Surv Circular 100:17–29

Ferrière L, Koeberl C, Libowitzky E, Reimold WU, Greshake A, Brandstätter F (2010) Ballen quartz and cristobalite in impactites: new investigations. In: Reimold WU, Gibson RL (eds) Large meteorite impacts and planetary evolution IV. Geological Society of America Special Paper 465, pp 609–618. doi:101130/2010.2465(29)

Fredriksson K, Dube A, Milton DJ, Balasundaram MS (1973) Lonar lake, India: an impact crater in basalt. Science 180:862–864

Fredriksson K, Brenner P, Dube A, Milton D, Mooring C, Nelen J (1979) Petrology, mineralogy, and distribution of Lonar (India) and lunar impact breccias and glasses. Smithsonian Contrib Earth Sci 22:1–12

French BM, Koeberl C (2010) The convincing identification of terrestrial meteorite impact structures: what works, what doesn't and why. Earth Sci Rev 98:123–170

French BM, Short NM (1968) Shock metamorphism of natural materials. Mono Book Corporation, Baltimore, 644p

French BM (1998) Traces of catastrophe: a handbook of shock-metamorphic effects in terrestrial meteoritic impact structures. LPI contribution no. 954, Lunar and Planetary Institute, Houston, p 120

Fudali RR, Milton DJ, Fredriksson K, Dube A (1980) Morphology of Lonar crater, India: comparison and implications. Moon Planets 23:493–515

Gasperini L, Alvisi F, Biasini G, Bonatti E, Longo G, Pipan M, Ravaioli M, Serra R (2007) A possible impact crater for the 1908 Tunguska event. Terra Nova 19:245–251

Grieve RAF (2005) Economic natural resource deposits at terrestrial impact structures. In: McDonald I, Boyce AJ, Butler IB, Herrington RJ, Polya DA (eds) Mineral deposits and earth evolution, vol 248. Geological Society of London Special Publication, London, pp 1–29

Grieve RAF, Langenhorst F, Stöffler D (1996) Shock metamorphism of quartz in nature and experiment: II. Significance in geoscience. Meteorit Planet Sci 31:6–35

Grieve RAF, Rupert J, Smith J, Therriault A (1995) The record of terrestrial impact cratering. GSA Today 5:189–196

Grieve RAF, Shoemaker EM (1994) The record of past impacts on Earth. In: Gehrels T (ed) Hazards due to comets and asteroids. University of Arizona, Tucson, pp 417–462

Grieve RAF, Robertson PB, Dence MR (1981) Constraints on the formation of ring impact structures, based on terrestrial data. In: Schultz PH, Merrill RB (eds) Multi-ring basins: formation and evolution, vol 12A, Proc Lunar Planet Sci. Pergamon, New York, pp 37–57

http://www.unb.ca/passc/ImpactDatabase/Age, 25 Apr 2011

Jourdan F, Moynier F, Koeberl C (2010) First $^{40}Ar/^{39}Ar$ age of the Lonar crater: a ~0.65 Ma impact event? 41th Lunar and Planetary Science Conference. The Woodlands, USA. Extended Abstract.1661

Kamo SL, Reimold WU, Krogh TE, Colliston WP (1996) A 2.023 Ga age for the Vredefort impact event and a first report of shock metamorphosed zircons in pseudotachylitic breccias and granophyre. Earth Planet Sci Lett 144:369–387

Kieffer SW, Schaal RB, Gibbons R, Hörz F, Milton DJ, Dube A (1976) Shocked basalt from Lonar impact crater, India, and experimental analogues. In: Proceedings of the 7th Lunar Science Conference. Lunar Planetary Institute, USA, pp 1391–1412

Koeberl C, Farley KA, Peucker-Ehrenbrink B, Sephton MA (2004) Geochemistry of the end-Permian extinction event in Austria and Italy: no evidence for an extraterrestrial component. Geology 32:1053–1056

Koeberl C, Reimold UW, Plescia J (2005) BP and Oasis impact structures, Libya: remote sensing and field studies. In: Koeberl C, Henkel H (eds) Impact tectonics. Springer, Berlin, pp 161–190

Koeberl C, Reimold WU (2005) Bosumtwi impact crater, Ghana (West Africa): an updated and revised geological map, with explanations, vol 145. Jahrbuch der Geologischen Bundesanstalt, Wien, pp 31–70 (Yearbook of the Austrian Geological Survey) (with map, scale 1:50,000)

Koeberl C, Martinez-Ruiz FC (2003) The stratigraphic record of impact events: a short review. In: Koeberl C, Martinez-Ruiz FC (eds) Impact markers in the stratigraphic record. Springer, Heidelberg/Berlin, pp 1–40

Kumar PS (2005) Structural effects of meteorite impact on basalt: evidence from Lonar crater, India. J Geophys Res 110. doi:10.1029/2005JB003662

Leroux H, Reimold WU, Koeberl C, Hornemann U, Doukhan J-C (1999) Experimental shock deformation in zircon: a transmission electron microscope study. Earth Planet Sci Lett 169:291–301

Mallet FR (1869) On the Vindhyan Series, as exhibited in the north-western and central province of India. Mem Geol Surv India 7(Pt. 1):129

Maloof AC, Stewart ST, Weiss BP, Soule SA, Swanson-Hysell NL, Louzada KL et al (2010) Geology of Lonar crater, India. Geol Soc Am Bull 122:109–126

Master S, Pandit MK (1999) New evidence for an impact origin of Ramgarh Structure, Rajasthan, India (abstract). Meteorit Planet Sci 34(suppl):A 79

Melosh HJ (1989) Impact cratering: a geologic process. Oxford University Press, New York, 245 p

Milton DJ, Barlow BC, Brett R, Brown AR, Glikson AY, Manwaring EA, Moss FJ, Sedmik ECE, Van Son J, Young GA (1972) Gosses Bluff impact structure, Australia. Science 175:1199–1207

Milton DJ, Glikson AY, Brett R (1996) Gosses Bluff – a latest Jurassic impact structure, central Australia. Part I: geological structure, stratigraphy, and origin. AGSO. J Austral Geol Geophys 16:453–486

Montanari A, Koeberl C (2000) Impact stratigraphy – the Italian record (Lecture Notes in Earth Sciences), vol 93. Springer, Heidelberg/Berlin, p 346

Murali AV, Zolensky ME, Blanchard DP (1987) Tektite-like bodies at Lonar crater, India: implications for the origin of tektites. J Geophys Res 92:E729–E735

Nayak VK (1972) Glassy objects (impact glasses), a possible new evidence for meteoritic origin of Lonar crater, Maharashtra State, India. Earth Planet Sci Lett 14:1–6

Nayak VK (1993) Maskelynite from the Indian impact crater at Lonar. J Geol Soc India 41:307–312

Osae S, Misra S, Koeberl C, Sengupta D (2005) Target rocks, impact glasses and melt rocks from the Lonar impact crater, India: petrography and geochemistry. Meteorit Planet Sci 40:1473–1492

Pati JK (2005) The Dhala structure, Bundelkhand Craton, central India – a new large Paleoproterozoic impact structure. Meteorit Planet Sci 40(Supplement):A121

Pati JK, Reimold WU (2007) Impact cratering a fundamental process in geoscience and planetary science. J Earth Sys Sci 116:81–96

Pati JK, Reimold WU, Koeberl C, Pati P (2008a) The Dhala structure, Bundelkhand craton, central India – eroded remnant of a large Paleoprotrozoic impact structure. Meteorit Planet Sci 43:1383–1398

Pati JK, Reimold WU, Koeberl C, Pati P (2008b) Dhala – a new, complex, Paleoproterozoic impact structure in central India, large meteoritic impact and planetary evolution IV, LPI contribution no. 1423, Lunar Planetary Institute, Houston, pp 168–169 (abstract)

Pati JK, Reimold WU, Pati P, Umrao RK, Ahmad M (2009) Study of pseudotachylite breccia vein, Dhala impact structure, India. 72nd annual meteoritical society meeting 2009, Nancy, (5187. pdf)

Pati JK, Jourdan F, Armstrong RA, Reimold WU, Prakash K (2010) First SHRIMP U-Pb and ^{40} Ar/^{39} Ar chronological results from impact melt breccias from the Paleoproterozoic Dhala impact structure, India. In: Reimold WU, Gibson RL (eds) Large meteorite impacts and planetary evolution IV. Geological Society of America Special Paper 465, pp 571–592. doi:101130/2010.2465(27)

Pilkington M, Grieve RAF (1992) The geophysical signature of terrestrial impact craters. Rev Geophys 30:161–181

Radhakrishnan BP (2005) Some thoughts on planetary impact structure in India and on the importance of their study. J Geol Soc India 66:423–431

Ramasamy SM (1987) Evolution of Ramgarh Dome, Rajasthan, India. Rec Geol Surv India 113 (7):13–22

Ramasamy SM (1995) Evolution of Ramgarh Dome, Rajasthan, India. Mem. Geol Soc India 31: 279–310

Reimold WU, Trepmann C, Simonson B (2006) Discussion – impact origin of the Ramgarh structure, Rajasthan: some new evidences, by Sisodia et al. 2006. J Geol Soc India, 68: 561–563

Reimold WU, Kelley SP, Sherlock SC, Henkel H, Koeberl C (2005) Laser argon dating of melt breccias from the Siljan impact structure, Sweden: implications for a possible relationship to Late Devonian extinction events. Meteorit Planet Sci 40:591–607

Reimold WU, Gibson RL (1996) Geology and evolution of the Vredefort impact structure, South Africa. J Afr Earth Sci 23:125–182

Sarangi S, Gopalan K, Kumar S (2004) Pb-Pb age of earliest megascopic eukaryotic alga-bearing Rohtas Formation, Vindhyan Supergroup, India: implications of Precambrian atmospheric oxygen evolution. Precambrian Res 132:107–121

Schultz PH, Merrill RB (1981) Multi-ring basins: formation and evolution. In: Proc Lunar Planet Sci, vol 12A. Pergamon, New York, 295 p

Sharma HS (1973) Ramgarh structure, India. Nature 242:39–40

Sisodia MS, Lashkari G, Bhandari N (2006) Impact origin of the Ramgarh structure, Rajasthan: some new evidences. J Geol Soc India 67:423–431

Spudis PD (1993) The geology of multi-ring impact basins: the Moon and other planets. Cambridge University Press, New York, 263 p

Stöffler D, Deutsch A, Avermann M, Bischoff L, Brockmeyer P, Buhl D, Lakomy R, Müller-Mohr V (1994) The formation of the Sudbury structure, Canada: toward a unified impact model. In: Dressler BO, Grieve RAF, Sharpton VL(eds) Large meteorite impacts and planetary evolution. Geological Society of America Special Paper 293, pp 303–318

Stoffler D, Langenhorst F (1994) Shock metamorphism of quartz in nature and experiment: I. Basic observation and theory. Meteoritics 29:155–181

Taylor SR (1992) Solar system evolution: a new perspective. Cambridge University Press, New York, 307 p

Chapter 16
Optical Characterization of Backscattering by Total Suspended Matter and Its Correlation with Phytoplankton Concentration in the Arabian Sea

Gunjan Motwani, Prakash Chauhan, Nivedita Sanwlani, and Hitesh Solanki

16.1 Introduction

The term total suspended matter broadly includes all the suspended particles in the water such as inorganic particulate material, white sands of coral beach and continental dust and clay. Action of water currents brings the sediments from the bottom into suspension. Further, the outflow of muddy rivers and estuaries, large tidal excursions, winds and volcanic eruptions also brings sediments into suspension. Presence of these particles significantly influences the ocean colour which varies depending on the dominance of the type of particulate matter in the water. Absorption $a(\lambda)$ and backscattering coefficients $b_b(\lambda)$ of suspended inorganic particulate material depends on various factors such as the geological structure and composition of the adjacent land, particle shape, size distribution and refractive index (Morel and Prieur 1977; Whitlock et al. 1981; Prieur and Sathyendranath 1981; Bukata et al. 1991; Gallie and Murtha 1992). The study of these inherent optical properties (IOP's) such as spectral absorption $a(\lambda)$, total scattering $b(\lambda)$, and backscattering $b_b(\lambda)$coefficients are essential for characterizing the marine optical environment and remote-sensing applications. In recent years, significant effort has been devoted to the development of models for estimating IOP's from underwater light-field measurements and apparent optical properties (AOPs) like irradiance reflectance $R(\lambda)$, remote-sensing reflectance $R_{rs}(\lambda)$ and diffuse attenuation coefficient for downwelling irradiance $K_d(\lambda)$ (Loisel and Stramski 2000; Gordon 1991; Kirk 1995; Roesler and Perry 1995; Gordon and Boynton 1997; Lee et al. 1998).

Studies on these optical properties of the sea are complicated because of their high variability with time and space. This paper reviews briefly the spectral

G. Motwani (✉) • H. Solanki
Department of Botany, School of Sciences, Gujarat University, Ahmedabad, India
e-mail: gunjan_motwani@yahoo.co.in

P. Chauhan • N. Sanwlani
Space Applications Centre, ISRO, Ahmedabad, India

R. Sinha and R. Ravindra (eds.), *Earth System Processes and Disaster Management*, Society of Earth Scientists Series 1, DOI 10.1007/978-3-642-28845-6_16,

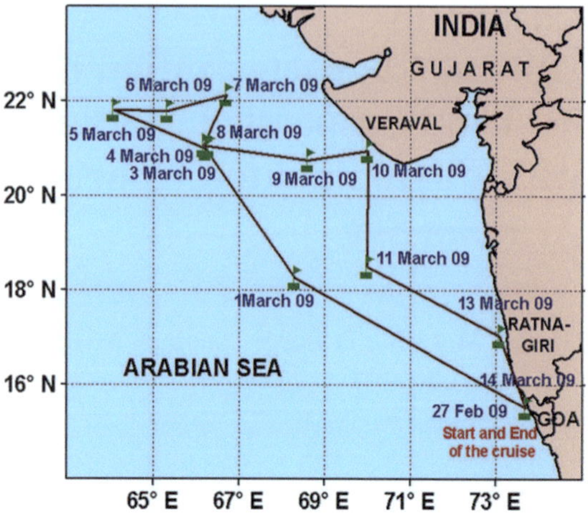

Fig. 16.1 Cruise track for SS-263

characteristics of backscattering behavior for total particulate matter and its correlation with chlorophyll-*a* (Chl-*a*) concentration over open and coastal waters of Northeastern Arabian Sea.

16.2 In-Situ Measurements and Data Collection

In-situ bio-optical measurements were conducted using SATLANTIC hyper-spectral under-water radiometer during 1st–13th, March 2009 onboard research vessel *Sagar-Sampada* covering wide variety of waters from Arabian Sea. Figure 16.1 shows geographical distribution of sampling locations and cruise track during the study. Upwelling water-leaving radiance (L_u), down-welling irradiance (E_d) and surface irradiance (E_s) measurements for a spectral range of 350–800 nm with 2 nm spectral interval were recorded along with Chl-*a* concentration backscattering coefficient (b_b) at 470 and 700 nm for 12 station with three casts each stations during this period. These have been used to compute the apparent optical properties like, Water-leaving radiance, $L_w(\lambda)$ and Remote Sensing Reflectance, $R_{rs}(\lambda)$ by use of following well-known equations (Fargion and Mueller 2000).

Water leaving radiance

$$L_w(0^+,\lambda) = L_u(0^-,\lambda)\frac{1-\rho(\lambda,\theta)}{\eta_w^2(\lambda)} \tag{16.1}$$

$\rho(\lambda,\theta)$ – Fresnel reflectance index of seawater

$\eta_w(\lambda)$ – Fresnel refractive index of seawater (1.345 for seawater is considered)

Normalized water leaving radiance

$$L_{wn} = L_w(\lambda)\frac{F_o(\lambda)}{E_s(\lambda)} \tag{16.2}$$

$F_0(\lambda)$ – The mean extraterrestrial solar irradiance
$E_S(\lambda)$ – Downwelling spectral irradiance at $z(0^+)$
$L_W(\lambda)$ – Upwelling radiance propagated through the surface $z(0^+)$

Remote sensing reflectance

$$R_{rs}(\lambda) = \frac{L_w(\lambda)}{E_s(\lambda, 0^+)} \tag{16.3}$$

$E_S(\lambda)$ – Downwelling spectral irradiance at $z(0^+)$
$L_W(\lambda)$ – Upwelling radiance propagated through the surface $z(0^+)$

16.3 Methodologies

Remote sensing reflectance (R_{rs}) acquired from in-situ measurements was plotted against the wavelengths ranging from 300 to 800 nm. This was done to study variation in R_{rs} from open and coastal waters at different wavelengths and thus facilitate comparison of optical components in those waters.

16.3.1 Chlorophyll-a Concentration Estimation Using Blue-Green Spectral Ratio Algorithms

Traditional blue-green ratios algorithms, based on $R_{rs}(412)/R_{rs}(554)$, $R_{rs}(442)/R_{rs}(554)$, $R_{rs}(490)/R_{rs}(554)$, $R_{rs}(510)/R_{rs}(554)$ and maximum spectral ratio out of ($R_{rs}(412)/R_{rs}(554)$, $R_{rs}(442)/R_{rs}(554)$, $R_{rs}(490)/R_{rs}(554)$, $R_{rs}(510)/R_{rs}(554)$) were computed in form of a statistical relationship to compare its variation with in-situ Chl-*a* concentration. Functional form of these algorithms has been tabulated in Table 16.1.

Table 16.1 Summary of the statistical results of Chl-*a* algorithms evaluation

Algorithms	Equation	Square of correlation coefficient (r^2)	Standard error
Spectral ratio ($R_{rs}412/R_{rs}554$)	$0.68*(R_{rs}412/R_{rs}554)^{-1.18}$	0.99	0.029
Spectral ratio ($R_{rs}442/R_{rs}554$)	$0.64*(R_{rs}442/R_{rs}554)^{-1.24}$	0.99	0.030
Spectral ratio ($R_{rs}490/R_{rs}554$)	$0.71*(R_{rs}490/554)^{-1.49}$	0.97	0.044
Spectral ratio ($R_{rs}512/R_{rs}554$)	$0.79*(R_{rs}512/554)^{-2.74}$	0.95	0.055
Maximum spectral ratio (($R_{rs}412 > R_{rs}444 > R_{rs}490 > R_{rs}510)/554$)	$0.72*(R_{max})^{-1.23}$	0.99	0.025

16.3.2 Angstrom Exponent Evaluation

Backscattering coefficient for suspended sediments was obtained through hyper spectral radiometer. Mean spectral slope for backscattering by particulate matter was then calculated in form of angstrom exponent of backscattering at 470 and 700 nm described as under.

$$\alpha(\lambda_1, \lambda_2) = \frac{\ln(\tau_a(\lambda_1)) - \ln(\tau_a(\lambda_2))}{\ln(\lambda_2) - \ln(\lambda_1)}$$

Here, α represents the Angstrom exponent (Angstrom 1964; Shaw et al. 1973), and the value of α depends on the ratio of small and large size particles.

16.4 Results and Discussion

Collected bio-optical data were plotted for analysis of their spectral properties. Figure 16.2 represents $R_{rs}(\lambda)$ spectra variation. Spectra exhibit maxima in green region corresponding to scattering by phytoplankton whereas minimal value in blue region corresponds to absorption by phytoplankton, Suspended Solids (SS), Coloured Dissolved Organic Matter (CDPOM) etc. A prominent fluorescence peak of phytoplankton in the wavelength domain 660–730 nm can be observed for coastal water spectra. In-situ measured Chl-*a* values were found to vary between 0.13 and 0.83 μg/L.

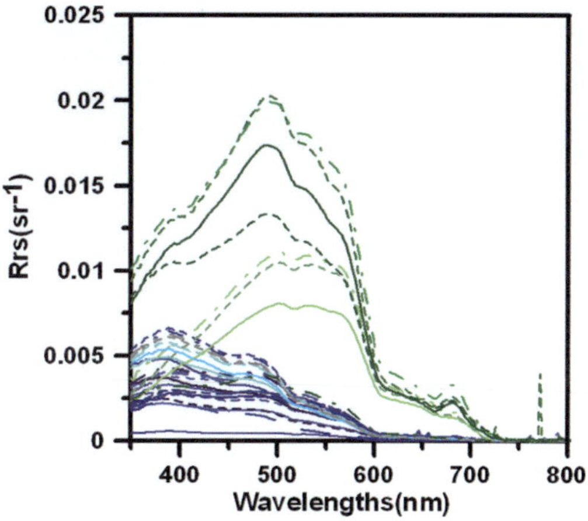

Fig. 16.2 Spectral variation of remote sensing reflectance (R_{rs}) for open ocean and coastal waters. The *blue lines* represent the ocean and the *green lines* represent the coastal waters

16.4.1 Blue-Green Spectral Ratio Algorithms

Comparative analysis was made between spectral ratio and in-situ measured Chl-*a* to get the best Chl-*a* algorithm. A summary of the statistical results of this comparison is shown in Table 16.1. Out of the mentioned spectral ratio algorithms, $R_{rs}(512)/R_{rs}(554)$ performed with $r^2 = 0.95$ as shown in Fig. 16.3. Comparatively better correlation was observed for $R_{rs}(412)/R_{rs}(554)$ and $R_{rs}(442)/R_{rs}(554)$ with same coefficient of determination $r^2 = 0.99$ and having root mean square (RMSE) = 0.029 and 0.030 respectively. In contrast maximum spectral ratio with $r^2 = 0.99$ and RMSE = 0.025 show better agreement with in-situ Chl-*a* concentration and were found to be superior to rest of the algorithms.

16.4.2 Correlation Between Chlorophyll-a and Backscattering Coefficient

Spectral variation of Backscattering coefficient can be observed from Fig. 16.4. The spectral slopes or angstrom exponent obtained was found to be −1.25 for open ocean waters and −0.49 for coastal waters as shown in Table 16.2. Higher values of α can be expected when smaller particles are dominant in comparison to larger particles. This shows that hygroscopic larger sized particles are present in Open Ocean waters. Phytoplankton cells alone dominate in this case. Whereas fine smaller sized sediment particles and phytoplankton both dominant together in coastal turbid waters and contribute in backscattering.

There exists a correlation between all the water quality parameters. Backscattering is highly correlated with chlorophyll-*a* concentration. A high correlation between backscattering and chlorophyll concentrations indicates that the optical properties of particulate assemblages are strongly influenced by phytoplankton and

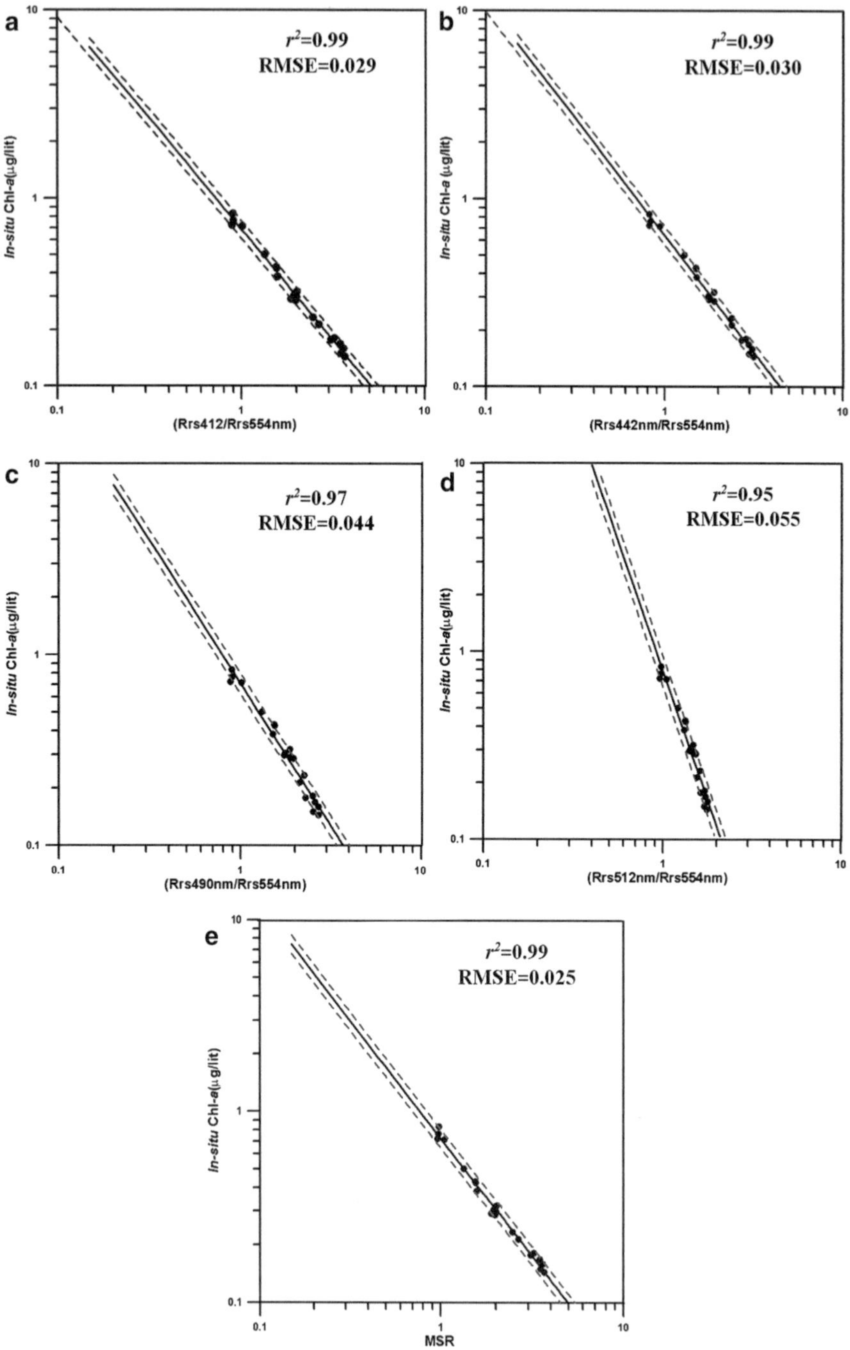

Fig. 16.3 Relationships between in-situ Chl-*a* concentrations (μg/L) and spectral ratios of (**a**) $R_{rs}412/R_{rs}554$, (**b**) $R_{rs}442/R_{rs}554$, (**c**) $R_{rs}490/R_{rs}554$, (**d**) $R_{rs}512/R_{rs}554$, (**e**) maximum spectral ratio (MSR). The *solid line* shows the best-fit regression. The *dotted lines* show ±5% error

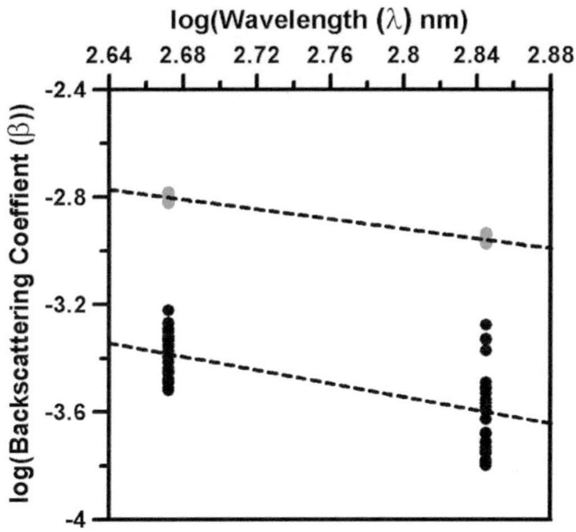

Fig. 16.4 Spectral variation of backscattering coefficients (b_b) in logarithmic scale

Table 16.2 Angstrom exponent in open and coastal waters

	Open ocean	Coastal waters
Formulation	AE = (log β_1–log β_2)/(log λ_1–log λ_2)	
Mean slope	−1.25	−0.49

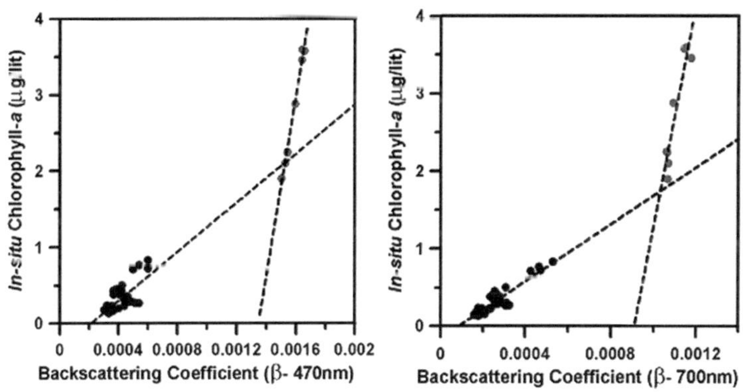

Fig. 16.5 Correlation between in-situ Chl-*a* and backscattering coefficients (b_b) at (**a**) 470 nm and (**b**) 700 nm

suspended sediments. The chlorophyll concentration estimated using maximum spectral ratio based chlorophyll algorithm was then correlated with backscattering coefficient at 470 and 700 nm as shown in Fig. 16.5. The results of this correlation are summarized in Table 16.3.

Table 16.3 Summary of results of Chl-*a* correlation with backscattering coefficient

		Open ocean	Coastal waters
470 nm	Equation	2276.7*β − 0.57	11854*β − 15.97
	Correlation	0.80	0.99
	Standard error	0.009	0.069
700 nm	Equation	1964.7*β − 0.16	11450*β − 13.7
	Correlation	0.93	0.88
	Standard error	0.056	0.272

Open ocean chlorophyll-*a* concentration shows a better correlation with backscattering coefficient at 700 nm with $r^2 = 0.93$ and RMSE = 0.056 whereas coastal turbid water shows a good correlation at 470 nm having $r^2 = 0.99$ and RMSE = 0.069. The backscattering component for wavelengths above 700 nm is very small for open ocean clear waters due to strong infrared absorption by water. Backscattering in open ocean waters is therefore dominated by phytoplankton (low concentration) alone and thus a good correlation is seen at 700 nm. Whereas presence of highly scattering sediments together with phytoplankton and yellow substance in coastal waters cause sufficient backscattering in NIR wavelengths and hence a high correlation at 470 nm. This diverse mixture of optical components makes the study of coastal water more challenging.

The task of developing a satisfactory, rigorous approach to address complexity of optical effects associated with the diversity of particle types and dissolved sea water constituents is extremely difficult. It is important to establish simple, empirical relationships, between the optical properties and the concentrations that can be applied immediately. Meaningful information can be derived from such simple relationships, for understand the processes that control the magnitude and variability of optical properties. These optical properties are critical for monitoring and detecting changes in ocean climatology and forecasting ocean biogeochemical processes.

16.5 Conclusions

Bio-optical measurements were carried out in the North-eastern Arabian Sea. These measurements were used for developing Chl-*a* algorithm using maximum spectral ratios of R_{rs} at 412/554, 442,554, 490/554 and 512/554. In-situ acquired backscattering coefficient at 470 and 700 nm was used to prepare a spectral plot and Angstrom exponent was computed form it. The Angstrom exponent was found to be higher for coastal waters. The Chl-*a* algorithm developed was used to estimate Chl-*a* concentration which was then correlated with backscattering coefficient. This correlation stood better for coastal water at 470 nm with $r^2 = 0.99$ and RMSE = 0.069 and at 700 nm with $r^2 = 0.93$ and RMSE = 0.056 for open ocean waters.

Such studies on backscattering coefficients and correlation with chlorophyll can provide important information about the dynamics of marine particles and dissolved organic carbon, because backscattering coefficients and chlorophyll are related to various biogeochemical parameters. Since chlorophyll is a function of phytoplankton, these correlations can be useful for developing novel primary productivity models and also in estimating phytoplankton growth rate, growth conditions and bio-mass.

Acknowledgements This work is supported under the Meteorology and Oceanography (MOP) II program of Indian Space Research Organization (ISRO). In this valuable advice and encouragement of Dr. Ajai, Group Director, MESG and Dr. J.S. Parihar, Deputy Director, RESA of Space Applications Center (ISRO), Ahmedabad is greatly acknowledged.

References

Angstrom A (1964) Techniques of determining the turbidity of the atmosphere. Tellus 13:214

Bukata RP, Jerome JH, Kondratyev KY, Pozdnyakov DV (1991) Satellite monitoring of optically-active components of inland waters: an essential input to regional climate change impact studies. J Great Lakes Res 17:470–478

Fargion GSY, Mueller JL (2000) Oceanoptical protocols for satellite ocean colour sensor validation, Revision 2. NASA Goddard Space Flight Centre, Greenbelt, MD

Gallie EA, Murtha PA (1992) Specific absorption and backscattering spectra for suspended minerals and chlorophyll *a* in Chilko Lake British Columbia. Remote Sens Environ 39:103–118

Gordon HR (1991) Absorption and scattering estimates from irradiance measurements: Monte Carlo simulations. Limnol Oceanogr 36:769–777

Gordon HR, Boynton GC (1997) Radiance-irradiance inversion algorithm for estimating the absorption and backscattering coefficients of natural waters: homogeneous waters. Appl Opt 36:2636–2641

Kirk JTO (1995) Estimation of the absorption and scattering coefficient of natural waters by use underwater irradiance measurements. Appl Opt 33:3276–3278

Lee ZP, Carder KL, Steward RG, Peacock TG, Davis CO, Patch JS (1998) An empirical ocean color algorithm for light absorption coefficients of optically deep waters. J Geophys Res 103 (27):967–978

Loisel H, Stramski D (2000) Estimation of the inherent optical properties of natural waters from irradiance attenuation coefficient and reflectance in the presence of Raman scattering. Appl Opt 39:3001–3011

Morel A, Prieur L (1977) Analysis of variations in ocean color. Limnol Oceanogr 22:709–722

Prieur L, Sathyendranath S (1981) An optical classification of coastal and oceanic waters based on the specific spectral absorption curves of phytoplankton pigments, dissolved organic matter, and other particulate materials. Limnol Oceanogr 26:671–689

Roesler CS, Perry MJ (1995) In situ phytoplankton absorption, fluorescence emission, and particulate backscattering spectra determined from reflectance. J Geophys Res 100 (13):274–294

Shaw GE, Regan JA, Herman BM (1973) Investigations of atmospheric extinction using direct solar radiation measurements made with a multiple wavelength radiometer. J Appl Meteorol 12:374–380

Whitlock CH, Poole LR, Ursy JW, Houghton WM, Witte WG, Morris WD, Gurganus EA (1981) Comparison of reflectance with backscatter for turbid waters. Appl Opt 20:517–522

Subject Index

R. Sinha and R. Ravindra (eds.), *Earth System Processes and Disaster Management*, Society of Earth Scientists Series 1, DOI 10.1007/978-3-642-28845-6,

MIX
Papier aus verantwortungsvollen Quellen
Paper from responsible sources
FSC® C105338

If you have any concerns about our products,
you can contact us on
ProductSafety@springernature.com

In case Publisher is established outside the EU,
the EU authorized representative is:
Springer Nature Customer Service Center GmbH
Europaplatz 3, 69115 Heidelberg, Germany

Printed by Libri Plureos GmbH
in Hamburg, Germany